Engelbert Kaempfer

EXOTIC PLEASURES

Fascicle III
Curious Scientific and Medical Observations

Translated with an Introduction and Commentary
by

Robert W. Carrubba

Published for the Library of Renaissance Humanism
by

Southern Illinois University Press
Carbondale and Edwardsville

Designed by P. B. Rollinson
Production supervised by Robyn Laur Clark

Frontispiece: Frontispiece of the *Amoenitates Exoticae* by Engelbert Kaempfer (Lemgo, 1712).

Illustrations from Engelbert Kaempfer's *Amoenitates Exoticae* are used with permission from Rare Books Division, Department of Rare Books and Special Collections, Princeton University Libraries.

Library of Congress Cataloging–in–Publication Data

Kaempfer, Engelbert, 1651–1716.
  [Observationes physico-medicae curiosae. English]
  Exotic pleasures : Fascicle III, Curious scientific and medical observations / Engelbert Kaempfer ; translation with an introduction and commentary by Robert W. Carrubba.
       p.   cm. — (Library of Renaissance humanism)
  Includes bibliographical references and index.
  1. Medicine, Oriental--Early works to 1800. 2. Science—Asia—Early works to 1800. I. Carrubba, Robert W. II. Title. III. Series.
R581.K3413 1996
    610' .95—dc20
    ISBN 0-8093-1976-4   94-29042
                                                                     CIP

The paper used in this publication meets the minimum requirements of American National Standard for Information Sciences—Permanence of Paper for Printed Library Materials, ANSI Z39.48–1984. ⊚

Requests for *Somnium et Vigilia in Somnium Scipionis (Commentary on the Dream of Scipio)* by Juan Luis Vives, edited and translated by Edward V. George, and for *On Poetry* by Giovanni Antonio Viperano, translated by Philip Rollinson, should be addressed to The Attic Press, Inc., 1502 Highway 246 North, Greenwood, South Carolina 29649. All other Library of Renaissance Humanism series titles can be obtained from Southern Illinois University Press, P. O. Box 3697, Carbondale, Illinois 62902-3697.

FOR
PETER AND CARLA CARRUBBA

Contents

Plates

Acknowledgments

I want to thank the Rare Books Division, Department of Rare Books and Special Collections, Princeton University Libraries for permission to reproduce the illustrations from Engelbert Kaempfer's *Amoenitates Exoticae*. My gratitude is expressed to Catherine Marraccini for her diligent proofreading. Special appreciation is due to Mary Kaye Adkins and Diane Locklin for the care and enthusiasm they devoted to the typing of the manuscript.

Introduction

Engelbert Kaempfer, the German scholar and physician, devoted an entire decade to cultural and scientific explorations. His travels began with his departure from Sweden in March 1683; they ended with his arrival in Holland in October 1693. The duration and scope of Kaempfer's journey through Russia, Persia, Arabia, India, the East Indies, Siam, and Japan complemented the breadth and depth of his curiosity and scholarship. Kaempfer was among the most learned scholars of his era, and he was certainly the most widely traveled. He has been called the Humboldt of the seventeenth century. One of his finest accolades came from the great botanist Linnaeus, who judged that Kaempfer more than any other man had earned the gratitude of the Japanese for bringing an accurate and sympathetic knowledge of this nation to the Western world.

Kaempfer was born in Lemgo, Westphalia, in 1651, just three years after the end of the Thirty Years' War, when Germany lay in ruins. Engelbert was the son of Johannes Kemper (the spelling of the family name was later altered), minister of Nikolai Church, and Christine Drepper, whose father had also served as pastor of the same Lutheran church. After his early years at the parsonage, Kaempfer studied at a variety of institutions in Germany and Poland, including the Latin schools of Lemgo and Hameln, the gymnasia of Lüneburg and Lübeck, and the athenaeum of Danzig. His 1673 dissertation at Danzig, *Exercitatio Politica de Majestatis Divisione*, reflects the climate of the absolute monarchies of Louis XIV in France and Charles XI in Sweden. Kaempfer's later education was taken at Thorn, at Cracow, where he studied languages, history, and medicine and earned a master's degree, and at Königsburg with emphasis on natural science and medicine, a happy blend of the liberal arts and sciences without which Kaempfer's reports and discoveries would not have been possible. Kaempfer possessed, indeed seems to have epitomized, virtues we ascribe to the Renaissance and the modern mind. With respect to the Renaissance, Kaempfer knew his Latin and Greek texts thoroughly and was a practicing researcher in both the humanities and the sciences. With respect to the modern mind, Kaempfer insisted on the primacy of reason and scientific methodology. These assertions, I believe, find solid confirmation in Fascicle III of *Amoenitates Exoticae,* or *Exotic Pleasures.*

Despite his splendid education, Kaempfer's career opportunities in Germany were not promising, while to the north lay Sweden with its university at Uppsala and an opportunity to study with the distinguished naturalist, Olof Rudbeck. The German's intellectual qualities impressed not only the faculty at Uppsala but King Charles XI of Sweden, who offered Kaempfer the position of court scholar. Kaempfer, however, now made a most significant decision. He chose instead to join the embassy of Charles XI to the Shah of Persia. As secretary of the Swedish Embassy, Kaempfer first traveled to Russia and the Court of Moscow, where he recorded a variety of observations, including an audience with the young tsars, Ivan V and Peter the Great. Before reaching Isfahan in March of 1684, Kaempfer survived a perilous crossing of the Caspian Sea. From this pre-Isfahan period comes Observation I, the Scythian lamb, a fabled plant-animal whose alleged existence Kaempfer rightly sought to disprove by firsthand investigation.

When the Embassy—whose charge included the establishment of commerce and the encouragement of a Persian break from the Ottoman Empire—had completed its negotiations at Isfahan, Kaempfer faced a second significant decision, probably the most important of his career. He was offered an appointment as court physician in Georgia, and he himself considered traveling to Egypt. But Father Raphael du Mans, a Capuchin and interpreter for the Persian court, persuaded Kaempfer to accept employment as a physician with the Dutch East India Company. It was during the period of travels in Persia from November 1683 through June 1688 that Kaempfer, when not at work or ill with a protracted fever, compiled materials for Observations II–VI: "the Torpedo of the Persian Gulf"; "Muminahi, or Native Persian Mummy"; "the Persian Dracunculus"; "Report of Disguun Asafetida"; and "Dsjerenang, or Dragon's Blood." Elsewhere in the *Amoenitates Exoticae* there are descriptions with splendid illus-

Introduction xvii

trations of Isfahan and Persepolis. Kaempfer saw Persia for the last time when he set sail from Gamron (now Bandar Abbas), the city that commanded the Straits of Hormuz. Under Admiral Lykochthon, the fleet weighed anchor, bound for Batavia, Java. It called at Dutch factories or trading stations in Arabia and in India, where Kaempfer gathered materials for Observations VII–X: "Andrum, or Hydrocele"; "Pericàl, or Ulcerous Hypersarcosis of the Feet"; "Snake Dances of Eastern India"; and "Two Indian Antidotes." In October 1689, the fleet dropped anchor at Batavia, the main Dutch settlement in the Far East. The city had been established in 1619 by Jan Pieterszoon Coen after he conquered the Javanese stronghold of Djakarta. In Java, Kaempfer continued his studies, but his eye was on the Dutch factory in Japan.

The events leading to the confinement of the Dutch on Dejima may be traced to 1542, when the Portuguese discovered Japan. During the next few years the lure of trade and profit drew seven Portuguese expeditions, until in 1549 Christianity made its way to Japan in the person of the saintly missionary, Francis Xavier. The rivalries of Portuguese and Spanish trade and missionary ambitions along with Japanese internal political and religious factionalism increased fears that the Westerners aimed at conquest, the first phase of which would be the establishment of a subversive Christian organization, an *imperium in imperio*. The total expulsion of both the Spanish in 1624 and the Portuguese in 1639 left the Dutch as the only intermediary between Japan and the Western world. Under a license from the Shogun, the Dutch established a trading post, called a "factory," on Hirado. The Dutch took pains to assure the Japanese that whereas the rival Portuguese were Catholics, the Dutch themselves belonged to a different sect of Christianity thoroughly opposed to the pope and his missionaries. In 1641 the Shogun instructed the Dutch to evacuate Hirado and move to the tiny, fan-shaped artificial island of Dejima, which had been constructed in 1635 in Nagasaki harbor. Rather than expel the Dutch and lose the advantage of trade and information concerning the Western world, the Japanese chose to confine one group of Europeans on Dejima.

The Dutch were not permitted to practice their religion openly or to observe the Lord's Day. To die on Dejima meant that one's body would find no repose in a cemetery but would, according to regulations, be committed to the waters about the island. Kaempfer summed up life on Dejima: "Thus we live all the year round little better than prisoners, confined within the compass of a small island, under the perpetual and narrow inspection of our keepers."

Yet there were opportunities to learn about Japan through interpreters and servants whom the European doctor paid, flattered, and instructed, and on the annual journey to Edo (now Tokyo) to offer homage to the Shogun and report on the state of the world. From Kaempfer's more than three years at Java, Siam, and Japan, we have most of Observations XI–XVI: "Acupuncture"; "Moxa"; "Japanese Tea"; "Ambergris"; "Persian and Indian Intoxicants"; and "Magic Spells of the Makassars."

Kaempfer returned to Holland in October 1693, more than ten years after he had left Stockholm on the mission to Isfahan. He assembled ten observations in the form of a dissertation that he presented to the faculty of medicine at the University of Leyden in April 1694, on which occasion he was awarded the degree of Doctor of Medicine. Nine of the ten topics of Kaempfer's *Disputatio Medica Inauguralis* reappear in Fascicle III of the *Amoenitates Exoticae* with expansion of scope and detail.

Instead of seeking an academic appointment, for which he was eminently qualified, Kaempfer returned in August 1694 to his place of birth, Lemgo. A demanding medical practice as physician to the Count of Lippe frustrated his intention of rearranging and polishing his vast store of materials for immediate publication. On 18 December 1700 he married Maria Sophia Wilstach, but their relationship proved less than harmonious. In the preface to *Amoenitates Exoticae* Kaempfer refers cryptically to "troublesome domestic problems." Maria Kaempfer bore three children, one son and two daughters, all of whom died in infancy. It was nineteen years after his return to Europe that Kaempfer finally saw the publication at Lemgo of his first major work, *Amoenitatum exoticarum politico-physico medicarum fasciculi V, quibus continentur variae relationes, observationes & descriptiones rerum Persicarum & Ulterioris Asiae.* The preface explains:

Unlike other travelers, I did not return laden with money and merchandise but with pages on which I had written exotic observations gathered through much labor, expense, and danger in various regions. I intended to publish immediately a number of volumes which were complete except for a few months of work on their organization and inter-relationships. Copper engravings also needed to be made, since exotica are very difficult to comprehend without the help of clarifying illustrations. A host of duties and problems confronted me, however, and led me astray from my consuming plan.

The topics of the five fascicles are: (1) the state of the Persian Court, (2) historical and scientific reports and observations on various things, (3) curious scientific and medical observations, (4) botanical and historical reports concerning the cultivation of the date palm in Persia, and (5) Japanese plants. The *Amoenitates Exoticae* was the last work published during Kaempfer's lifetime, for after several attacks of colic, he died at Lemgo in 1716 at the age of sixty-five.

Shortly before his death Kaempfer completed his *History of Japan*. Sir Hans Sloane (1660–1752), president of the Royal College of Physicians from 1719 to 1735, an avid collector of specimens of natural history, books, and manuscripts, learned from the introduction to the *Amoenitates Exoticae* that Kaempfer's manuscripts and collections were in Germany and arranged for their purchase. The manuscripts included *The History of Japan*, which was translated from the German into English by Sir Hans's Swiss librarian, Johannes Casparus Scheuchzer, and published in 1727. The Sloane collection of Kaempfer's materials is now housed in the British Museum.

EXOTIC PLEASURES
Fascicle III

Observation I
The Scythian Lamb, or the
Borometz Fruit

The credulity of the ancients created the phoenix from descriptions of the date palm tree.[1] In a similar fashion, the curiosity of a more recent era was again deceived by terminology. This era transformed an animal into a vegetable, and thereby introduced the vegetable or plant, the Borometz or Barannetz,[2] into the histories of the exotic.[3] It is so pleasant to believe in the existence of the Borometz that Jean Bauhin, the great prince of the realm of plants, seems to have envied the distinguished Scaliger the glory from the first report. After listing those who recorded this plant before Scaliger, Bauhin[4] says: "To be sure, the report of so great a wonder merited the author the reward of honor; but that very great and unrivaled man was so eager to know all or to be thought to know all that even in this matter he relied solely on himself and was unwilling to take a step toward his own glory with the aid of another's name."[5] The plant, Bauhin writes, grows in Tartaria (Baron von Herberstein[6] locates it in Zavolha;[7] others place it around the Caspian Sea) in the shape of a lamb to a height of about three feet, is covered with a very fine skin (which the inhabitants remove and use for head coverings), feeds on grass, and is devoured by wolves.[8] It serves no purpose to repeat the remaining details; they are all accurate and well known to the reader. But the reader must understand that these details apply to a real animal (a lamb or lamb fetus) from the same land of Uzbek or Tartaria Minor[9] and not to any species of vegetable. I did not learn this truth from careless inquiry but from experience, the true teacher.[10] I will now present my findings for the open-minded reader.

Baran is a word of Slavonic origin, which in Polish and Russian means sheep. Its diminutive in Polish is Barannek; in the Russian spoken at Moscow, Borannetz; and by corruption, Borometz. Not unlike Baran is the Persian word of the same meaning, Barreh, which is also common among the neighboring Scythians. There are, however, two breeds of Scythian sheep.

The first breed is the ordinary one familiar in Germany. The second breed[11] is unique and is found in certain provinces around the Caspian Sea such as Chorasmia,[12] Bokhara[13] and Nagaja.[14] This second breed is remarkable in the followinq respects:

1. *Size.* The breed so far surpasses the normal size of other breeds that frequently its members are as tall as asses, and all are much taller than three feet.

2. *Tail.* Its tail is different from the tail of our sheep. These sheep have on their hind parts an enormous mass of fat that leaves a trail and is so heavy that sometimes the sheep are unable to carry it about. To say that the tail sometimes weighs forty pounds may invite disbelief but it is a fact.

3. *Meat.* Its flesh is more sought after than that of other animals. The fat, especially that supplied by the massive tail, is exceedingly delicious and sweet. Because of its excellence, the fat is used in place of butter for cooking rice and all varieties of food.

4. *Hide.* The hide is truly exceptional and is particularly valued for the fineness and elegance of the curled hair.

Now let us return to the issue in question. The use of these hides is common among both the Scythian and Persian peoples, since in accordance with Mohammedan religious law they abstain from garments of other hides on the grounds that these are obtained from impure animals or animals suspected of impurity. But the nobility and the wealthy in their pride desire clothing beyond the lot of the ordinary man. They long for the small skins of young animals, which are far more delicate than the older ones. The younger the little lamb, the more costly is the skin removed from it. Furriers work the wool into finer and tighter curls, which make the little skins attractive and expensive. And so it happens that men bent on softness and money-making have no qualms about anticipating the birth of the lambs with an act of barbaric butchery. Solely for the sake of obtaining the little skins, they cut open the womb of the mother while she is still pregnant. After proper preparation, a small skin is of such dubious nature and so exceedingly delicate that when the extremities are cut off, it scarcely bears any resemblance to lambskin. An uninformed and credulous person can mistakenly believe it to be some type of downy membrane of the gourd.

This is the method of preparation. The inner surface is cleansed of impurities and outer cuticle by the usual technique. The wool on the outer surface is curled again and again; this process, however, will not succeed unless accomplished in sunlight and in the following manner. Two men grasp the little skin and vigorously stretch it. The wool is combed to remove the flock. Then it is worked with a brush (Bürste) and continuously turned in circles. All the while, a supervisor keeps sprinkling water held in his mouth. The energetic working, together with the alternate wetting and drying, reduces the wool into tight and tiny curls, which will soon be rendered firm and flat by the application of a gentle press.

The price of a little skin soars to three or more gold pieces according to its estimated quality. The little skin is used for trimming turbans and frequently serves as an ornamental border for robes and cloaks. A skin acquired by a Caesarean operation is called a Kalema; a skin stripped from an animal already born is called Mimrass. Both terms are employed among furriers, but because they are unknown to the general public, I did not wish to write them in the Persian alphabet.

The Arabian desert toward Basra[15] sustains the same breed of sheep, which is equal in size to the Scythian. There is, however, less esteem and use for the skins in this region that neither practices nor knows the cruel act of disemboweling the fetus. Further, all lower Asia as far as Palestine sustains this breed of sheep, which is covered with a different type of hide and does not achieve the gigantic proportions of the Scythian sheep. It does, however, furnish equally sweet fat. The people of the East spread it on their foods and especially on rice. The fat gives these foods a taste far more gratifying than any expected of butter or any other fat.

On these grounds, I assert the following. Inasmuch as there is no knowledge or record anywhere in Tartaria among both the ordinary people and expert botanists (as I investigated to my humiliation and disgust) of the existence of a zoophyte[16] that feeds on grass; and inasmuch as nothing called Borometz (except for a herd of sheep) can be found in this area, whatever is reported about this plant is pure fiction and fable. Perhaps the Borometz was created by the speculations of some philosopher. Perhaps it was created by the ignorance of the first one to make the report, who through a faulty knowledge of the language or occasional inattentiveness misunderstood what he was told. Perhaps we should ascribe it to some other circumstance whereby the little skin with its enigmatic delicateness was carried through far distant lands and thus lost its true history and accurate name. At length, the little skin arrived among us as a charming prodigy and came to the attention of the distinguished Scaliger, a man of curiosity and an admirer of this foreign fur. With its vegetable-like appearance, it easily, as do all things marvelous, gained credence as a wonder. And so this error, dignified by Scaliger's authority and soon affirmed in books, so captivated the minds of the most intelligent men and the opinion of the public that in this very day there is displayed in museums among the rarities as a species of zoophyte something which is beyond a shadow of doubt the small skin of a fetus obtained by a Caesarean.[17]Barclay wisely says, "Reported prodigies are usually received with favor and admiration, and although they do not conform to the truth, after they have once found an author, they are pleasing to many, grow in veneration and are esteemed for their antiquity."

Notes

1. This chapter is a more precise account of the first topic of Kaempfer's doctoral dissertation, *Disputatio medica inauguralis exhibens decadem observationum exoticarum* (Leyden, 1694). Kaempfer discusses the *Palma dactylifera* fully in *Amoenitates Exoticae*, 659–764. The Egyptian word *bennu* means both the date palm tree and the phoenix. Similarly, the Greek word *phoinix* stands for the date palm and the mythological bird. The origin of the connection between tree and bird is uncertain. The myth of the Scythian lamb was widely believed in Europe from the thirteenth century on. It was publicized in England by Sir John Mandeville (fl. 1356) in *The Voyage and Travels of Sir John Mandeville, Knight*.

2. Also spelled *Barometz* or *Borametz*.

3. The myth also inspired a number of poems. Dr. Erasmus Darwin (1731–1802), grandfather of Charles Robert Darwin, wrote in *The Botanic Garden* (1792):

E'en round the Pole the flames of love aspire,
And icy bosoms feel the secret fire,
Cradled in snow, and fanned by Arctic air,
Shines, gentle Borametz, thy golden hair;
Rooted in earth, each cloven foot descends,
And round and round her flexile neck she bends,
Crops the grey coral moss, and hoary thyme,
Or laps with rosy tongue the melting rime;
Eyes with mute tenderness her distant dam,
And seems to bleat—a "vegetable lamb."

4. Jean Bauhin (1541–1613, botanist and physician), *Historia plantarum universalis nova et absolutissima* (Yverdon, 1650–51), 406.

5. Julius Caesar Scaliger (1484–1558) contributed to botany, zoology, and especially literary scholarship. Bauhin objects to Scaliger's omission of previous reports. Scaliger, however, in his *Exotericarum exercitationum liber XV* (Frankfurt, 1557), 181.29, is discussing the report of the plant-animal by Gerolamo Cardano (Jerome Cardan, 1501–1576) in *De rerum natura* (Nürnberg, 1557), 6.22. Curiously, Scaliger's sarcastic remarks were interpreted not as a rejection of Cardano's account but as a confirmation. Scaliger thus became the champion of the fable of the Scythian lamb.

6. Sigismund von Herberstein, *Rerum muscoviticarum commentarii* (Vi-

enna, 1549). This is a work of prime importance for the history of Russia.

7. Zavolha is a region between the Volga and Ural rivers.

8. In *Pseudodoxia epidemica* (1646) Sir Thomas Browne (1605–1682) wrote, "Much wonder is made of the Boramez, that strange plant-animal or vegetable Lamb of Tartary, which Wolves delight to feed on, which hath the shape of a Lamb, affordeth a bloody juyce upon breaking, and liveth while the plants be consumed about it."

9. A region in south central Asia now part of the former U.S.S.R.

10. A point that deserves to be underlined. Kaempfer, as subsequent chapters demonstrate, wrote not from hearsay but on the firm base of his own careful and accurate observations, often the result of arduous and dangerous travel.

11. The Karakul.

12. Kaempfer lists Chorasmia as one of the five main divisions of Persia. It was the eastern part, extending to India, and was itself divided into 40 provinces (134–35).

13. A part of Uzbek and formerly a state.

14. A region west of the Caspian Sea.

15. A city and port of the Shatt al Arab about sixty miles from the Persian Gulf.

16. A plant-animal. The term *zoophyte* (now obsolete) covered plants supposed to possess qualities of animals, and occurs in English as early as 1621.

17. In *The Vegetable Lamb of Tartary: A Curious Fable of the Cotton Plant* (London, 1887), Henry Lee contends that Kaempfer's "small skin" is in no way connected with the origin of the tale of the Barometz (30). Even so, there remains the question of the museum exhibits Kaempfer mentions. Were some of these in fact the dried and shrunken skins of the real lamb Kaempfer describes? Lee remarks, "Kaempfer's suggestions were ingenious, though his theory was erroneous" (24). Lee concludes that a version of the myth, which mentions the existence of a tree that bears fruit or pods that ripen and burst open to reveal little lambs, points to the cotton pod (*Gossypium herbaceum*) as the real source. Compare Lee's illustrations 3 and 61. Hence the similarity of cotton and wool, and the misunderstanding of such figurative phrases as "fleece that grows on trees" would account for the origin of the myth. For a brief discussion, see Willy Ley, *Exotic Zoology* (New York, 1966), 58–61.

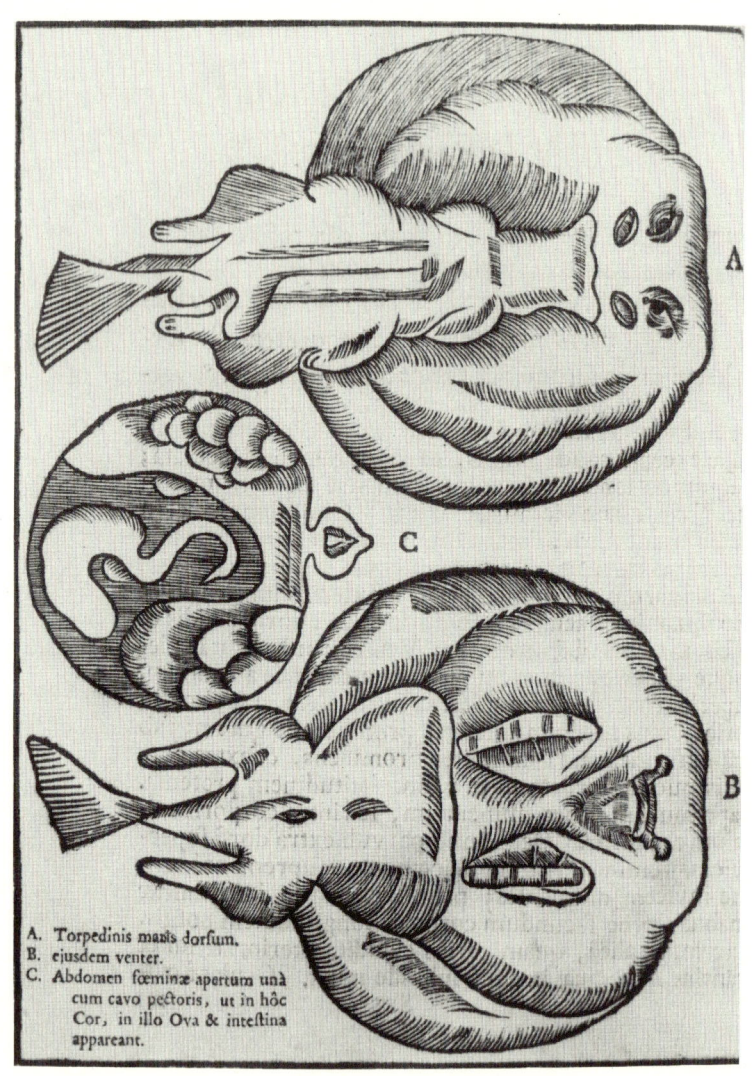

A. Torpedinis maris dorsum.
B. ejusdem venter.
C. Abdomen fœminæ apertum unà
    cum cavo pectoris, ut in hôc
    Cor, in illo Ova & intestina
    appareant.

Plate 1. *Torpedo*

Observation II
The Torpedo of the Persian Gulf

I
Anatomy

As in the Latin language, the names of the torpedo fish in Persian and in Arabic are derived from the numbness that it inflicts upon anyone who touches it.[1] The Persian name is lers mahii; the Arabic is riaad. Rich in fish with scales, the Persian Gulf supports an abundance of torpedoes which, especially during the winter months, enter unwanted into the nets of fishermen. The shape of the torpedo's body is compressed. Except for the tail, the torpedo resembles the ray, but its shape is more nearly circular. The largest torpedo I saw did not exceed eighteen inches in diameter.[2] The average torpedo is two inches thick at the middle and gradually thins out toward the extremities. The substance of the middle is cartilaginous; that of the extremities is softer, boneless, and suits the requirements of a fin. The torpedo's exterior is slippery and has no scales. It is marked by a variety of spots: white and dark spots on the back, less distinct ones on the tail. The torpedo's stomach is white as is usual with fish fond of the surface and the depths of the sea. The torpedo has an uneven surface on both its sides. This is especially true of the back, whose center swells in the shape of a little shield from beneath which rises a distinctly prominent tail. The tail also extends more than the width of a hand beyond the circumference of the body. The head does not protrude; on the contrary, it is enclosed within the circle of the body. Tiny eyes protrude from the surface of the back. The width of a thumb separates the eyes from one another and from the outer edge of the body. Each eye is equipped with two eyelids positioned nearly on line with the long axis of the back.

The outer eyelid, which the torpedo rarely closes, is sturdy; the inner eyelid, with which the torpedo repeatedly blinks in the water, is thin and clear. Behind the eyes at about the width of a straw are located two spiracles or openings.[3] They are positioned obliquely with

9

reference to the eyes, whose size they approximate. While the torpedo
is in water, it frequently covers the openings with a very tender skin
the way a man winks. A casual observer would think the torpedo has
a second pair of eyes; perhaps this is just what happened to the
distinguished Borrichius.[4] The mouth is found on the belly opposite
the location of the eyes. The size of the mouth is such that although
when closed it can be covered with a segment of the thumb, nevertheless by the lateral turning of the lips it may be stretched into a huge
opening. Where the lips are depressed into the oral cavity, they bristle
with very sharp and very fine spines so situated that whatever is taken
in will hardly be able to get out. After these spines comes an array of
very sharp teeth, which are widely spaced and hidden far back on the
jaws. At both folds of the mouth lie crescent-shaped curves or openings
like noses. In the interior these are separated from the oral cavity, a
soft craticle (the type concealed beneath the jaws of fish). Stretched
over and supported by a bony partition, they do not collapse. Small
openings mark the middle region of the stomach (a part, of course, of
the abdomen), which is exceedingly spongy, soft, and very thin. Five
openings on each side are arranged in rows; they are narrow, not at
all long, and are positioned transversely. Each opening is closed by a
very strong piece of skin. The nerves correspond to the openings in
length, positioning, and arrangement in rows. The anus has an oblong
opening in the part of the stomach where the tail extends beyond the
circular form of the body. Under pressure, the anus emits black and
earthy feces along the white and very thin worms a palm in length.

The tail is thick, cone-shaped, and ends in an obliquely rising fin
with a decussate extremity. This fin is preceded by two other fins
similarly positioned in one row but at slight intervals. The two fins
have ovate extremities. The somewhat larger of the two fins is closer
to the middle of the back; the smaller fin is closer to the end of the
tail. The initial portion of the tail is accompanied on both sides by a
somewhat larger, fleshy fin, which expands into a flat semicircle two
inches in diameter. On the males, both these fins end in a one-inch
quasi-phallus,[5] which is cartilaginous, thin, striated in one way or
another, and has two tiny openings at the tip. A slight pressure expels

from these tiny openings a milky, thick, and viscous liquid. A disemboweled torpedo furnishes its dissector with taut skin, bluish white flesh, strong peritoneum, cartilaginous vertebrae running from the back to the cone of the tail, no spines but in place of them tendons coming from the vertebrae.

The cerebrum has five prominent pairs of nerves: the first pair extends to the eyes, the last accompanies the medulla for a modest distance, the others soon run in various directions from their source. The heart, whose shape is precisely that of a fig, is freely suspended in the small pectoral cavity. The abdomen has a capacious stomach strengthened by many fibers and filled with black, fetid excrement. The stomach has a number of veins; one stands out among the rest. As it extends to the right lobe of the liver, it touches the vesicle containing bile. On the left of the orifice of the stomach there is a small, oblong, bluish, irregularly shaped gland, which might have been the pancreas or the spleen. The liver is a large, pale red organ composed of two lobes. One lobe fills the entire right side of the cavity; the other occupies the left side of the cavity but is smaller and more compressed with a prominent vein swollen with dark blood. One would identify this second lobe as the spleen, except that it is united with the other lobe by a thin strip (below the chest) and is clearly of the same substance and color. The lobes are filled with many adherent glands, which may be embryonal growths, and when dissected, they pour out a rich fluid that resembles butter. When the intestines and stomach have been removed, a certain thin vessel attached to the back on both sides is disclosed. The vessel is clear, curved, uneven, and intricate with many winding ducts.[6] To it adheres in one way or another a certain fleshy substance not unlike what anatomists call bat's wings. One might have said the vessel was a uterus or an ovary. On the female I found a very large number of eggs resting on both lobes of the liver. The eggs were not enclosed by a shield but by a thin skin of a pale, sulfurous color, and they resembled perfectly the yolks of chicken eggs. The eggs floated in a clear mucous liquid and were enclosed by a common ovary, a clear, tender, and tiny membrane joined with the liver. As the oppressive weather did not permit me to remain indoors for long, I was forced to forego a more detailed anatomy. I shall,

however, review the wondrous faculties that the Persian torpedo
exhibited and that I observed in the open air: faculties that have been
occasionally and uncritically described by scientists or reported by
other people.

II
Faculties

Our Persian torpedo appears to differ in important respects from
the torpedo caught in the Mediterranean Sea, if the latter has been
correctly described by Aristotle,[7] Pliny,[8] and Galen.[9] Perhaps the
difference is the same kind evident in the case of the tarantula: in Persia
the tarantula never causes with its venom those tragedies that the
tarantula in Italy is reported to stage by Kircher,[10] that venerable
mystic of nature, and quite recently by the distinguished Baglivi.[11]
The torpedo does not, to be sure, constantly exhale its horrifying
vapors; at intervals and by a voluntary effort it explodes when as it
attempts to flee it senses its freedom obstructed or is annoyed by being
handled. The torpedo emits its power with a sort of momentary
belching or a certain convulsive motion of the viscera, whereby it
dilates the spiracles of the abdomen and absorbs air; with the same
effort it simultaneously thrusts out its dreadful virus into the air. With
a quite similar thrust but with a less relaxed bodily motion the
porcupines of Africa and the Orient eject spines at those annoying
them,[12] and certain petulant apes suddenly begin to shake in order to
strike terror into anyone touching them. When grasped in water, the
torpedo's strike is less powerful, either because the water intercepts
the force or because in its proper element the torpedo is not thoroughly provoked. Usually the torpedo may be touched for a short
time out of water without any injury to the hands until, either
impatient with the air or annoyed by the pressure, it suddenly discharges its virus. When freshly removed from the sea, it strikes more
frequently and more sharply; it strikes less frequently and potently if
kept and handled for several hours. To be sure, the livelier, the larger,
and the more recently captured torpedo numbs more powerfully and
frequently. Again, the female torpedo appeared to me to strike with

more effect than the male.

When handled, the torpedo strikes hardest at the arms and shoulders; similarly, when annoyed by a foot (even one protected by a shoe), it directs its dreadful numbing force chiefly at the knees, shins, and thighs. Complaints about increased palpitation of the heart come more from those struck on the foot than from those struck on the hands. Those who had once or twice experienced the numbness claimed to feel vaguely the horrifying chill of the vibrating torpedo simply by moving a hand close to it; no such claim is made by those who for the first time approach the torpedo and have not yet been rendered numb by contact. Fishermen deny that a torpedo caught in a net can be felt by drawing on the ropes; certainly the numbness is not conducted to the hands of a person provoking the fish with a staff nor to the hands of a person touching it with a spear or rod, as Pliny writes[13] and scholars believe. The numbness induced is not the sort felt in a sleeping limb, but a sudden condition that instantly travels through the touching part and penetrates the citadel of life and breath. Then it overwhelms the whole body and mind, as it seizes the sinewy and bony parts such as the hands, shins, and elbows. In a word, you would think that your major joints were broken and limp, especially those on that member that first received the expelled vapors. And all of this is accompanied by a shudder of the heart, a trembling of the limbs, a numbness, and a chill. So powerful and so swift is the force of the horrifying exhalation that like a chill bolt of lightning it shoots through the handler.[14] No reward could induce a person (provided the torpedo is not languishing close to death) to endure the force of the exhalation with sufficient ease so as not to cast the fish away immediately.

While I was handling torpedoes, I had an experience worth reporting. A certain African from the group of spectators boldly lifted up the torpedo repeatedly and held it without any sensation of horror. I asked him to explain how he did this. He said, "Take a deep breath filling the lungs and hold it, being careful not to exhale. For as long as you can manage this, you will feel no harm from the fish." I tested his method with success. Others to whom I revealed this new discovery also succeeded in avoiding harm by holding their breath. This is my explanation. As the initial step in the process of retaining air, we rather fully exhaled our breath from the body in all directions; this drove off the miasma emanating from the fish. Now after the breath had been held for a rather long time and the lungs were allowed to relax considerably, we seemed again to sense faintly the poisonous exhalations from the fish.

The torpedo's life is fragile. If kept in an ample container filled with seawater, the torpedo will not live until the next day. If handled briefly during the morning, it will be dead in the afternoon. A dead torpedo may be handled without any injury. The natives even claim that it is edible, but as a precaution against its numbing effect a captured torpedo is released. Fishermen believe that nature has endowed the torpedo with the ability to induce numbness as a weapon with which it can render aggressors and neighbors powerless and torpid.[15] Aristotle asserts this,[16] Pliny agrees,[17] and swift anchovies, which I have on a number of occasions found along with other small fish in the stomachs of torpedoes, are proof. When placed along with other fish in a cask filled with water, the torpedo did not demonstrate this, perhaps because in captivity it is inclined to ignore an enemy. No use is made of the torpedo: if captured, it is released. For this reason I was able to obtain them from fishermen for a small price. The distinguished Ludolf reports that the Abyssinians cure fevers by applying a torpedo (caught in rivers and lakes).[18] The efforts of Sennert[19] and others are misdirected in their search for a cure for the numbness discussed. The condition ceases spontaneously and instantly with no damage to the body or humors. Mattioli prints an illustration of a female Italian torpedo.[20] It is very similar to our Persian torpedo except for the pattern of spots and the tail fin, which on our fish is erect and decussate; the tail fin on the other fish is round and expands on the same plane as the body.

Notes

1. Torpedo comes from the Latin *torpere*, to be stiff, numb, or torpid. The names "numbfish" and "crampfish" also demonstrate selection of name from the effect produced. In *De Natura Deorum*, Cicero writes: "tutantur . . . torpore torpedines," that is, "torpedoes protect themselves with torpor"(2.50.127). While at the port of Gamron, Kaempfer had an opportunity to study first hand the varieties of fish caught in the Persian Gulf. His attention was caught by a species of torpedo which, while it resembled the Mediterranean torpedo, differed anatomically and with respect to its faculties as they had been reported by ancient and more recent authors. The Mediterranean torpedo, called *Torpedo marmorata* Risso, is in fact a different species from the Persian Gulf torpedo. Kaempfer is thought to have described *Torpedo Sinus-Persicus* Olfers, which is now usually placed in synonymy with *Torpedo panthera* Olfers. I am grateful to Dr. Eugenie Clark, of the University of Maryland, for her generous assistance in identifying Kaempfer's torpedo. See Harald Blegvad, *Fishes of the Iranian Gulf* (Copenhagen, 1944) pt. 3, 44 and plate 3, fig. 1. For a photograph of the *Torpedo panthera* (mottled electric ray) see K. Kuronuma and Y. Abe, *Fishes of Kuwait* (Tokyo, 1972), 45.

2. About twenty species of the electric ray (family *Torpedinidae*) are known to come from warm waters. Some species achieve a weight of 200 pounds. On the east coast of the United States is found the *Torpedo nobiliana*; on the west coast, the *Torpedo californica.*

3. The spiracles or large holes lead to the pharynx and take in water for respiration.

4. Olaus Borrichius (1626–1690), Danish physician. Borrichius, as Kaempfer says, made the mistake of thinking the torpedo had a second pair of eyes.

5. The posterior ends of the pelvic fins are modified to form copulatory organs.

6. This is the electric organ of the torpedo. The dorsal surface is positive, the ventral negative. For a discussion of this organ see Harry Grundfest, "Electric Fishes," *Scientific American* 203 (1960): 115–24, 220.

7. References to the torpedo are numerous in the works of Aristotle (384–322 B.C.). For a useful discussion of ancient sources see D'Arcy Wentworth Thompson, *A Glossary of Greek Fishes* (London, 1947), 169–171.

8. Pliny the Elder (Gaius Plinius Secundus, A.D. 23–79), a man of encyclo-

pedic interests. For the torpedo see *Historia Naturalis* 9.57, 78, 143, 162, 165; 32.7, 94, 102, 105, 133, 135, 139, 151.

9. Galen (c. A.D. 129–199), a friend of the emperor Marcus Aurelius and one of the most famous physicians of antiquity. For his references to the torpedo see C. G. Kuhn, *Claudii Galeni Opera Omnia* (1821–33; facs. Hildesheim, 1965) 20:525 .

10. The bite of the tarantula was thought to induce tarantism, a nervous disorder characterized by hysteria and an uncontrollable desire to dance (tarantella). Tarantism, the dancing mania, originated in the fourteenth century in Apulia and spread throughout Italy, reaching its zenith in the seventeenth century. In *De arte magnetica*, 3rd ed. (Rome, 1654), Athanasius Kircher (1601–1680) defends the theory, which Kaempfer rightly rejects (593).

11. Giorgio Baglivi (1668–1707), Italian physician.

12. In actuality, porcupine quills may be erected at will but they cannot be thrown.

13. Pliny maintains that the torpedo is effective at a distance or by contact with a spear or rod (*Historia Naturalis* 32.7).

14. This well-chosen comparison between the power of the torpedo and that of lightning is the closest Kaempfer comes to identifying the true element operative. Benjamin Franklin's celebrated experiment with kite and key during a thunderstorm took place in 1752, forty years after the publication of Kaempfer's book. W. Cameron Walker comments, "This hypothesis of effluvia or corpuscular action was supported to some extent by Kaempfer, who appears to have been the earliest writer to compare the effect with that of lightning and therefore the first to link the phenomenon with electricity" ("Animal Electricity before Galvani," *Annals of Science*, 2 [1937]: 88).

15. The torpedo voluntarily emits its electric shock for defensive and offensive purposes—that is, against both predators and prey.

16. Aristotle, *Historia Animalium* 620b19.

17. Pliny tells us that the torpedo hides in the mud and snaps up fish that have received a shock while swimming carelessly above it (*Historia Naturalis* 9.143).

18. In *Historia Aethiopica* (Frankfurt, 1681) Hiob Ludolf (Leutholf) records the medical employment of the torpedo on humans for fevers, gout, and exorcism. The procedure was in a sense an early form of electric shock treatment: the patient was strapped to a table, and the torpedo was applied to various parts of the body (1.11).

19. Daniel Sennert (1572–1637), German physician noted for early descriptions of scarlet fever, scurvy, and dysentery.

20. Pietro Andrea Mattioli, *Commentarii in libros sex Pedacii Dioscoridis Anazarbei de medica materia, Opera quae extant omnia* (Frankfurt, 1598), 257.

Observation III
Muminahi,
or
Native Persian Mummy

I
Native Mummy in General

Name. The Persians prefer a certain native liquid with wondrous powers to the local pearls and precious stones that are showpieces of their treasures.[1] The Persians term the liquid Muminahi,[2] which is its proper name and means Mummy. They also call it either Belessoon, which is a general term of lofty significance and means Balsam, or Kodreti, which is a laudatory title and denotes a truly free gift from God and nature: natural mummy.

In my judgment it is worthwhile for me to present for our country an account of this substance, which a few travelers have investigated but no one has ever described.[3] There is no need to mention that this most estimable remedy has never reached our apothecaries. The reason, of course, is that this substance, which drips very sparingly from hard rock, is reserved as a sacred and exclusive possession of the royal palace and is employed within the circle of the most serene family. Only on rare occasions do small amounts reach the nobles through the kindness of the King, if he learns that someone has fallen from his horse (as happens during a hunt) and is critically injured. In the opinion of the people, this is the genuine and authentic mummy of the ancients with which the early Egyptians preserved the bodies of their leaders[4] with the result that our people termed the bodies themselves mummies. The same term was then retained for the artificial balsam that is made from spices and is used as a substitute for genuine mummy in preserving corpses. Consequently, the Persians refused to apply the term Kodreti to so-called Egyptian mummy on the grounds that it was unworthy of this most exalted title; instead, they employed the term Ensaani, a word of Arabic origin that means something human. Or the Persians might have recourse to the substantive Mum which means wax or the substantive Muminahi which means the liquid balsam here discussed and authentic mummy. Their intention was, of course, to distinguish their own genuine mummy from the adulterated mummy of the Egyptians.

Description. The mummy, which I propose to discuss, is a bituminous liquid that exudes from the surface rock of a mountain. It has the foul appearance, color, thickness, and something like the quasi-viscosity of cobbler's pitch. While the mummy is fresh and adheres to rock, it has a greater fluidity. When heated, it is tractable, mixes readily with oil but not with water, is quite odorless, and is very similar in substance to Egyptian mummy. When placed over live coals, it emits the strong smell of sulphur, which has been tempered somewhat by naphtha and is not entirely unpleasant to the nose. The odor resembles that which is produced through suffumigation by either dry naphtha, or asphalt, or ancient Egyptian mummy, or even dark amber. For all these are bitumens whose substances are by no means unrelated; they differ from one another by degrees of strong odor or sweetness due to the varying disposition and amount of sulphur or, if you prefer, of their own salt.

II
Report of Collection

There are two types of mummy. The first type is primary mummy, which is precious because of its superb powers and scarcity. We here express the highest praise for it. The second type is secondary mummy, which nature bestows in greater abundance but endows with a lesser effectiveness. Secondary mummy is normally employed as a substitute for primary mummy and is sold fraudulently as the powerful primary mummy. The source of primary mummy is very remote from man's travels, villages, and water supplies. It is an especially deserted site in the province of Darab, one day's journey from the city of Dara.[5] Dara was named after its founder, Darius, who was the last Persian monarch and the one quite famous because of his celebrated conquests.[6] The mummy is found here in a certain cave that resembles a well as it descends two fathoms into rock at the foot of a steep mountain of the Caucasus. Now the Caucasus, rising in Iberia, divide Asia, in the words of Curtius, with an unbroken ridge.[7] Scattered branches wander all about Persia; they are barren in most places, but in this area the mountains are extraordinarily rugged and wild.

Collection of the mummy is said to have been interrupted for several centuries, either because the flow ceased or because the location was lost in the turbulence of wars and buried in forgetfulness. The location was rediscovered and collection again undertaken at the beginning of the seventeenth century after Christ. Since that time, the mummy has been gathered annually with great pomp and ceremony. In order to ensure the authenticity of this precious liquid, which is destined for a place among the treasures of the royal house, the highest administrators of the provinces are themselves responsible for collecting it at the source. The time of collection is set for that part of the summer when the mummy softens the most from the intense heat of the Dog Star and with minimal effort comes free from the uneven wall of rock.

The ritual and technique of collection are as follows. The supreme prefect of the provinces of Lar[8] and Darab meets with the other royal assistants from both regions at the appointed time at the foot of the mountain. They inspect and remove the seal with which they marked the cave the previous year. Then the huge rock that sealed the entrance is rolled away by twenty sturdy laborers, and one man enters to strip clean the bitumen. He is equipped with an iron spoon that is appropriately beaked for scraping the walls. In order to deny the collector any opportunity for theft by concealing mummy in his clothing while he is alone and unobserved, he removes his clothing and is sent in nude except for the privy parts. To preclude theft by swallowing, the collector is given clear water to hold in his cheeks until he emerges. Having entered the cave, he scrapes off whatever mummy has come forth from the rock during the course of the previous year. This task takes him about one hour. Meanwhile, the deputies, who will spend the time in feasting, return to the magnificent tent that was pitched for this purpose in the area of the cave. Upon returning from the pit, the collector gives the mummy to the inspectors and discharges into a silver dish the water that he had kept in his mouth so that all may determine whether he has substituted his urine for the water he was given. Indeed, he is led from the tent and fingers are inserted into his anus to determine whether he has hidden any of the precious liquid in his bowels. Then and there the mummy is liquefied over a fire so that the useless bits of rock scraped off along with the bitumen may settle to the bottom. The clear matter is poured from the plate into a small silver box newly made for this purpose. The usual yield is twenty-five mescals,[9] that is, a bit more than four ounces (since a mescal equals a drachma plus a few grains). After the box has been sealed by the five leaders of the assembled group, it is entrusted to a swift courier for transport without delay to the royal palace at Isfahan.[10] The five officials in charge divide among themselves the remaining impure portion—a practice winked at by the King. Finally, after the entrance to the cave has once again been closed and sealed, the assembly is dismissed.

Secondary Mummy

Secondary mummy exudes in small quantities from rocks in the same stretch of desert mountains between the cities of Lar[11] and Darab. Since nature does not produce it with the same zeal, secondary mummy is far less noble and is inferior in power. For this reason, if it is discovered, secondary mummy may be collected by anyone provided he is able to do so despite a steep descent. The most famous type of secondary mummy and the one most frequently substituted for the precious primary mummy is Sehebbenaad. The rustics, who collect this mummy, so named it after the place where it is found.[12] It adheres to a high and steep ridge from which the man who scrapes it off is lowered by a rope—a terrifying spectacle. From below archers also dislodge pieces by shooting arrows. Its smoke resembles that of sulphur raw with naphtha and is heavier than the smoke of precious mummy. An owner offered to sell some to me for an equal quantity of silver by weight. Another secondary mummy, which was given the rustic name of Tsjenpeh or Tsjampeh from its obscure location,[13] was on sale for four times its weight in silver. Tsjampeh mummy is free of the raw, heavy odor. Its smoke resembles that of amber black with asphalt. It has a sweetness very close to or perhaps even greater than that of the most noble mummy of ancient Egypt, whose friability it also possesses.

By the term ancient mummy I do not mean that foul mixture that is on display in the shops and goes by this name. Nor do I mean the human bodies that have dried in the desert sands and are commonly called mummies. What I do mean is the famous balsam of the antiquity with which the Arabs and Egyptians filled the bodies of the higher nobility. I find two types. The first type, very precious and exceedingly rare, was extracted from the corpses of the nobles of western Asia. It attests itself genuine Darab mummy by the likeness of its substance, by preciousness, and by the sameness of name. Its smoke, however (which rather resembles that of Tsjampeh mummy), is evidence that it has been mixed with benzoin, styrax, or opobalsam. Perhaps as a result (unless this is due to age) it tends somewhat to turn red when crumbled. The second type, which we Europeans carefully preserve in abundance and without discrimination as the authentic ancient mummy, filled humbler bodies and consequently betrays the cheaper matter from which it was produced. I am prepared to believe that this second type is asphalt, whose quite genuine scent it emits.

I have seen a liquid very similar to native mummy pouring forth on a certain mountain of the naphthiferous peninsula of the Caspian Sea. The mountain is located three miles from the Median city of Baku.[14] As the liquid flows from the summit, it is rather fluid at first and gradually congeals as it makes its way down. The ignorant rustics only use it for fire to heat bath water. But black naphtha, which is drawn from wells about a mile away, constantly flows over the earth because of careless handling and hardens into a very similar resin. This resin differs from the famous secondary mummy because it has a heavier odor from its sulphur, which also emits fumes of black naphtha. When fresh, the resin need not even be burned to emit this odor, but with age the odor disappears. Of the same character and quality is the dry, congealed bitumen of Strabo (*Geography* 16), whose source he adduces from Eratosthenes to be liquid bitumen or naphtha from the nearby Babylonian plain.[15] I have not seen this location, but

I suspect that even in our time the dry bitumen can be found there, since this is where the liquid is obtained. So great is the parity of the aforementioned mummies with dry naphtha, asphalt, amber, and ancient mummy in substance, color, and consistency, so great is their affinity in odor and power, that I scarcely refrain from asserting that all are bitumens of the same type. Which of these happens to have a greater or lesser sweetness of odor and excellence of power depends partly on the differences in filtration by the place of origin and partly on the differences in intensity of and burning by subterranean fire. Consequently, either by percolation or refining each obtains a slightly different purification of its sulphur and disposition of particles.

III
Powers and Uses

Now let us return to primary mummy. Primary mummy possesses every virtue in which ancient mummy, whether Arabian or Egyptian, is recorded to have excelled. It is so superbly effective for uniting bones that it is believed to make the fractured limbs of animals firm and strong enough to use within several days, provided the bones are properly reduced. In the case of offspring, in fact, the period of healing is three days; for young fowl, that very same day. The effectiveness of this medicine has become so celebrated among the people of Persia and is believed with such faith that to entertain any doubts about it is in their eyes a sacrilege.[16]

The mummy is employed both internally and externally. Internally, it furnishes a superbly effective surgical balsam for healing abscesses and ulcers, for dissolving blood clots, and for ruptures and other injuries that harm the internal organs in cases of accidental falls from high places. A small amount of mummy, liquefied with a sufficient quantity of butter, should be imbibed. The patient must carefully avoid wetting his teeth with it, for they say that it harms and loosens teeth. This, they believe, is the distinguishing mark of the nobler unadulterated mummy. Externally, mummy is effective for dislocated bones. After proper reduction the bones are anointed only once with the aforementioned liquid compound in place of a plaster.

In fact, if the bones are anointed in this way before being reduced, the tendons, they claim, are so contracted after one night that the dislocation cannot subsequently be reset by any technique. The best results, as I said, are obtained in very serious bone fractures by both internal and external use. A small amount of mummy is liquefied in a sufficient quantity of butter. Part of the liquid, containing about five grains of mummy, is given to the patient to swallow; the remainder is smeared on a linen cloth and fastened to the properly reduced fracture. After this ligature has once been applied, the bandages are not removed until the limb can be used. When the leg of a young fowl has been broken, two or three grains are adequate for both uses.

Experiment. As is my custom with wonders of nature, I did not hesitate to class this astonishing effectiveness with other claims whose truth must stand the test of personal examination and experience. Hence, I experimented with fowl, not once but as often I happened to obtain some mummy either by purchase or as a gift. In the course of these experiments I followed all instructions with precision. Since all my efforts to obtain primary mummy proved fruitless, I substituted secondary mummy or the mummy that was being fraudulently sold as primary mummy. It is said to possess the same powers but to act more slowly. What were the results? There was always some quite laudable effect, but it was far short of my desires and the boasts of the mummy's supporters, for I discovered that the bones of the fowl did not heal much more quickly with the bitumen bandage than without it.

As chance had it, some time later at a gathering of nobles I was berating as excessive the claims made for the mummy. A member of the household of the governor of Lar immediately came forward to defend the mummy against my mistrust. He intended to prove the excellence of royal mummy (a small amount of which he was fortunate to possess) for uniting bone before, as he said, he would dismiss me. He approached the matter with great confidence before a crowd of spectators. I here record the whole affair for the reader's judgment. He took a quantity of precious Darabic mummy the size of a small chickpea (two or three grains), and to it he added three times the amount of secondary mummy for a larger subject. He ordered this to be liquefied with a little butter (about a drachma) in a silver spoon that I supplied for the task. While this was being done, a six-month old fowl was brought out and I was assigned the job of breaking either of its legs. I completely broke one leg so that the bone protruded from the broken skin. I reduced the fracture, placed a warm band smeared with our balsam around it, and bandaged it firmly. The remainder of the liquefied balsam I poured down its gullet. Afterwards I kept the fowl in a small, dark place. All of this was carried out in accordance with the instructions of the mummy's champion. Those who had been present on the previous day as sponsors and spectators assembled at the precise hour on the following day. The fowl was brought forth from the darkness and set free after its bandages had been removed. The fowl spread its wings, flapped them in joy over its freedom, and ran off spryly in search of food but with an ever so slight limp such as might be brought on even in a healthy foot by the pressure of the bandages. Soon it gave evidence that it was completely free from pain by dashing for the grain scattered about and by the fact that it boldly attacked other fowl in an attempt to drive them off from the food. Those of us who had doubted the mummy's powers stood in awe.

When the sponsors demanded that I endorse their position on the mummy's effectiveness I replied, "Unless some carpenter glued the fracture, there is certainly no natural way for it to be joined so soon. Even if I witnessed and acknowledged the extraordinary power of the mummy, I would still not believe that nature, assisted by even the strongest remedy, could within a day accomplish so great a task in a bloodless leg: to secrete the substance for callus, to unite and to harden."

Without delay, I dissected the leg and made a visual examination of the fracture. I found that the outer skin covered by the balsam ointment was much thicker than usual and was firmly binding the fractured bones. After removing the skin with a scalpel, I also discovered an amazingly thickened periosteum, which, by wrapping the fracture like a bandage, held and strengthened the bones in their proper position. Once this membrane was scraped away, however, there was no question concerning the belief that the fractured bones united by intermediate callus. Quite the contrary: a gentle touch separated the bones. Instead of the first signs of union, a slight amount of blood was visible on the edges of the fractured bone. Hence, it was clear that nature had admirably begun her work but had not perfected it. Because of my confidence in this experiment, I subsequently tested the power of secondary mummy (I had purchased a supply of it while traveling through the region where it is produced) on a human subject. I also obtained callus more quickly than with any porotic,[17] but on the whole it grew beyond the desired amount.

Notes

1. Asians and Europeans alike firmly believed in the power of mummy, although they might not have agreed on the nature of the substance and on its effects. Over a period of centuries a curious confusion arose about just what mummy was. Originally mummies were sought after because it was believed that they were prepared with natural bitumen, a substance highly regarded for its medicinal value. As time passed, people forgot that the embalmed bodies were prized not for themselves but for the bitumen or asphalt that they supposedly contained so that eventually the therapeutic value was ascribed to the mummies themselves. When the supply of authentic mummies could not satisty the demand, unscrupulous dealers created spurious mummies from the bodies of criminals and slaves; see Warren R. Dawson, "Mummy as a Drug," *Proceedings of the Royal Society of Medicine* 21 (1927): 34–39. Ultimately the term *mummy* could mean nothing more specific than medicated flesh, whether human or animal. Nicholas Culpeper prescribes, "Take an Owl, pull off her feathers, and pull out her guts, salt her well for a week; then put her into a pot and stop it close, and put her into an oven. So she may be brought into Mummy" (*Culpeper's Last Legacy* [London, 1671], 77). And with regard to the beneficial effects of mummy, the claims were many and conflicting. Mummy was reputed to be an antidote for poisons, to correct disorders of the liver, to aid the heart and lungs, to cure ulcers, to dissolve blood clots, to heal internal injuries, and to unite bone fractures. Arab and European physicians prescribed the drug for both internal and external uses, and dispensatories included mummy as a normal entry. It is worth noting that as early as the sixteenth century the great surgeon Ambroise Paré in his *Discours de la Mumie et de la Licôrne* (Paris, 1582) pointed out the uncertain nature of mummy and condemned the drug as worthless. Paré not only refused to prescribe mummy himself, but when serving as a consultant, he urged other physicians not to expose their patients to it (see especially chapter VIII).

2. Also written as *mumiyeh* or *mumiai.*

3. For travelers' reports on mummy, see George N. Curzon, *Persia and the Persian Question* (London, 1892), 2:521.

4. The ancient geographer Strabo (64 B.C.–A.D. 21) asserts that the Egyptians used a bitumen for embalming their dead (*Geography* 16.2.45), but modern research has rejected this much-repeated contention. Warren R. Dawson concludes,

> In general terms, it may be said that for the immersion-bath common salt (mixed with various impurities), and not natron, was used. For the sub-

sequent anointing, the principal ingredient was juniper-resin. The resins of several coniferous trees have been identified, and cedar and olive oil were also used. The presence of alcohol in some of the tissues lends support to Herodotus' statement that palm-oil was used for cleansing. Crude natron, as mentioned before, was often sprinkled upon the body after treating it with resin as a dehydrating agent. The resin, which was used in large quantities in mummies of late periods, was applied in a molten condition, and in this state often closely simulates both pitch and bitumen, but must not be confounded with these substances as is usually done; there is as yet no evidence that bitumen was ever used. With regard to the packing materials, apart from rags of linen which formed the exclusive material used for filling the body-cavity in the Eighteenth and Nineteenth Dynasties, lichen, sawdust, sand and mud were used in the Twentieth and succeeding dynasties, and sometimes a mixture of butter and soda was used for packing the face. ("Making a Mummy," *Journal of Egyptian Archaeology* 13 [1927]: 40–49)

5. Darab is a district of the province of Lars. The town of Dara, or Darab as it is usually called, is the capital of the district and is located 140 miles southeast of the city of Shiraz.

6. Legend ascribes the foundation of Darab to Darius III (c. 380–330 B.C.). The original name *Darabgerd* means "Dariustown." Darius was defeated by Alexander the Great at the battles of Issus and Gaugamela. His subjects then deposed Darius and stabbed him to death at the approach of Alexander.

7. Quintus Curtius Rufus (fl. c. A.D. 70), *History of Alexander* 7.3.19. Kaempfer is following the ancient historian's concept of one long chain of mountains stretching from the Mediterranean to India.

8. The district of Lar is located directly south of Darab.

9. Also spelled *miscal* or *miskal*, a Persian and Arabian measure of weight.

10. In 1604 Isfahan was made the capital of Persia by Shah Abbas I.

11. The city of Lar, the capital of Laristan, stands at the foot of a mountain range in a large plain covered with palm trees. Lar is 180 miles southeast of Shiraz and 75 miles from the coastal city of Lingeh.

12. Behbehan is located northwest of Shiraz and not far inland from the head of the Persian Gulf. Mummy seeps from cracks in the limestone core of the Kuh-i-Khaiz where there is a gorge cut by the Marun River.

13. I am unable to identify this "obscure location."

14. For Kaempfer's illustrated description of Baku on the Apsheron Peninsula, see *Amoenitates Exoticae*, 268ff.

15. Strabo, *Geography* 16.244. Eratosthenes of Cyrene (c. 275–194 B.C.) was the head of the Alexandrian Library and the most versatile scholar of his era. Recognized as the first systematic geographer, Eratosthenes calculated with considerable accuracy the circumference of the earth and the size of the moon and the sun.

16. In the *Canon of Medicine* the celebrated Persian physician Avicenna (980–1037) prescribes mummy for a wide variety of ailments, including fractures, concussions, paralysis, and disorders of the liver and stomach.

17. porotic: a medicine supposed to produce callus.

Observation IV
The Persian *Dracunculus*
on the Coast of the Persian Gulf

I
Name

The way of life led by the inhabitants of Hormuz, or rather of the whole coast of the Persian Gulf, is very hard: they suffer constant bodily misfortunes from the extreme weather. Now they drip with perspiration, now they are troubled with boils; presently they lack food, soon they lack water; often they suffer from burning winds, always they suffer from the scorching sun. Who could list all their misfortunes? I present one of their more serious hardships: a unique type of worm that is bred not in the intestines but in the muscles on the exterior of the body. The Latin physicians call the worm the *dracunculus*,[1] either because of a certain resemblance to a snake, or because it lodges coiled in the muscles like a serpent, or because it moves under the skin in the manner of a serpent—if we believe the account of Rhazes, the tenth-century physician. Sennert (the German Galen)[2] and a number of distinguished writers confuse the *dracunculus* with *crinones puerorum*[3] or the excretions from the pores of the body, which due to their viscosity look like little worms, and are called Mitesser[4] in German. Avicenna used the term Irk Medini which means Medina nerve. Avicenna's translator, Gerard of Cremona, missed the exact meaning of Irk either through carelessness or lack of knowledge, and incorrectly rendered the term Irk Medini as Medina vein (*Canon* 4. 3. 2. 21).[5] By Medina is meant the town in Arabia, which on account of its excellence the Arabians simply call the city or the city of the prophet, because when Mohammed fled from Mecca he made Medina his home. The blacks of African Guinea call the worms Ickon in their language, as one learns from natives brought from the Gold Coast. The Persians on the northern shores of the Persian Gulf term these foster children Pejuuk and Naru. But we must not delay over names.

The question of names has been quite fully treated by Velschius, a physician of broad research and erudition (*Exercitat. d. Ven. Med.*).[6] I propose to relate only a few things about the *dracunculus* and its attributes—things that I have gathered by my personal examination and experience.

II
Description

I confidently call the *dracunculus* a worm, even against the eyewitness of Avicenna. Because he had doubts as to whether or not the dracunculus was a living being, Avicenna preferred to label it a nerve or a vein. Usually, the *dracunculus* is extracted over a period of time or is drawn out a segment at a time along with pus. When casually examined under such conditions, the dracunculus clearly has somewhat the appearance of a nerve. However, I seem to have been more fortunate in assisting the issue of the *dracunculus*, for on two occasions the uncommon happened: with one effort, I withdrew a *dracunculus* from the cavity of the scrotum. The first patient was an African; the second, a Persian. When brought into the light of day, both worms displayed spontaneous motion and life. This superb demonstration was observed by witnesses (of whom two were young surgeons) invited by me to confirm the facts. It was noteworthy that: when placed in warm water, it immediately became limp, weak, and motionless; when taken out, it at once hardened and moved, although not very distinctly; but when immersed in cold water, it curved itself vigorously in one direction or another and, being submerged and impatient of the cold water, it repeatedly raised its head above the surface of the water. Its movement was a sure indication of the sensation of pain—clearly different than the motion observed when a cord or thread becomes wet and coils out of necessity and with a mechanical positioning of its parts. Consequently, we physicians who never believed our patients when they complained of a live and moving subcutaneous worm that caused sudden pain now have abandoned the opinion of Ebn Sina [Avicenna]. We will henceforth believe that there exists a real living being, a species of worm, which the greatest of the Arabs, Avicenna, mistakenly called a vein, as do Zacuto Lusitano[7] and the very keen observer, Mercuriale.[8]

The worm is pure white in color, and of an uncertain length (a foot, a cubit, or more). It is tender and terete, with about the thinness of a (thicker than average) lute string. The worm is equipped with a beak, which the Persians call a Riisj, or a beard; indeed, under microscopic examination, it is said to resemble hair. The beak is white, pellucid, a finger in length, hair-thin, and so delicate that, when drawn out into the air or extracted with an epithem, it droops of its own accord and escapes the notice of unwary surgeons. At the root of the beard, one can see a dark point and even the trace of a mouth along with an unprotected eye; at the end of the tail, one can see the point of an opening—the indication of an anus. The worm has a double tunic; both are whitish. The inner tunic is a little more delicate, and its cavity is filled with a white, rich, and sometimes serous fluid but not even a tiny drop of blood may be seen.

III
Generation and Birth

This little beast is native not only to Persia but also to African Guinea and (witness Rhazes and Avicenna) to Chorasmia[9] and Egypt. The much lauded Avicenna informs us that the region of Medina in Arabia abounds in the little beasts and he treats the malady under the title of its country (Medina Irk) in chapters 21 and 22, *Canon* 4.3.2. I have discovered that certain inland villages in the land of the Ganges beyond Paliacatta[10] also produce *dracunculi*. I also found traces in the Tartarian desert near the Ural River where it flows into the Caspian Sea. However, I also made the following discoveries. Every land that produces *dracunculi* is hot and parched from an intemperate climate. Its soil is salty, barren, sandy, and neither supplies water to the surface nor maintains any in its depths, unless the water is tainted with brackishness and unfit for quenching thirst. Consequently, the winter rains (very sparse especially in the region of Hormuz), which gather on the surface of the land, are collected in large cisterns painstakingly constructed for this purpose. This rain water is zealously conserved as the annual supply of drinking water granted to the people by the indulgence of a none-too-generous heaven for their sojourn in this land. Future researchers on the generative process of insects should take due note of this. The worm makes its appearance in that part of the summer beset by the glowing Lesser Dog Star—as it would only be fitting for the unlucky beast to be born under an evil omen. Summers differ in intensity of heat, and I have discovered that the hotter the summer, the more productive it is of worms. The worm makes its home in the leg rather than in other parts of the body, especially in the portion of the leg from the knee to the very sole of the foot. It is less frequently bred in the thigh, from the loins down to the knee. Its most usual locations are the internal and external malleoli. Occasionally, I have removed worms from the sole of the foot and from the arms. In all, I have once or twice withdrawn a worm hidden under the kneecap, in the joint of the knee and in the scrotum, or from the hand, the hip, the loins, and the side. To the best of my memory, I have never removed a worm from any other location. Usually, the *dracunculus* ranges widely and is concealed by a swelling of the flesh: at one time, it is coiled in many circles above a muscle; at another time, it is stretched straight out; at still another time, it is fixed in the interstices of the muscles; often, it is strikingly entwined about the ligaments of the limbs; most often, it winds its way tortuously through the muscles. This is perfectly clear from the pain, movement, and the ulceration that affect various locations on the body and from the fact that the removal of a single *dracunculus* may involve various locations on the body. I removed another worm from a clerk over a period of three weeks. While I was drawing out one end, which was existing from beneath the knee, the great toe on the patient's foot, as if pulled by a continuous string, experienced strong pain. Eventually, the other end of the worm was secreted from beneath the toe along with a large quantity of pus. I removed another worm from a carpenter: first, one end came out from the top of the calf, the middle portion came out from beneath the ankle, and finally the other end, which extended to the sole of the foot, exited with very great difficulty. Enough said on the location of the worm; a few comments now on the worm's origin and related matters.

IV
Emergence

A slight fever of one day, sometimes lasting for three days, accompanied by a slight inflammation and swelling of the area containing the worm usually indicates that the worm is about to emerge. On the next day, a pustule arises on the location; the pustule is as large as a pea, tender, and clear or occasionally dark.

When after one or two days the pustule has broken spontaneously or has been pricked, the initial part, the tiny beak, which was concealed inside, comes forth. The rest of the body gradually follows from various places when induced daily by gentle tension. Often with only a preliminary pain and a scarcely perceptible induration the *dracunculus* emerges. With the breaking of the ulcer, the worm makes an uninterrupted path for itself without causing a pustule of the epidermis. The excretion of the worm is the easier or the more difficult, the quicker or the slower, according to its size and rate of maturation as well as situation and location. Rarely is the worm extracted whole and living; most frequently the extraction requires a period of time, about ten days. A worm lodged in the cavity of the scrotum makes the most felicitous exit, for after its beard emerged I removed a live worm from this location without the pain and inconvenience of a festering ulcer. A worm, which is not fully extended and lies or is coiled over a muscle of the leg or arm (where it can be seen and felt), can be extracted within a few days after maturation begins, without notable pus or annoyance; in fact, its extraction may on occasion be accomplished with a single effort. Generally, a worm located in the thigh or elbow comes forth easily within a very few days. When a worm whose location is unsettled envelops the ligaments of the foot, its extraction entails serious difficulties over a period of twenty or more days and is accompanied by a very large daily discharge of corrupt matter and great pain and trouble for the patient. This happens if the worm is improperly grasped and broken. The worm breaks and reenters, if it is securely entwined about or fastened to ligaments, especially at a point quite distant from the point of exit. The worm does not mature as a whole, but part by part, so that as one end (the head, to be sure) emerges, the rest of the body is to all appearances still being created and produced as by apposition (a term used in the schools). In such circumstances, success in extracting the worm varies considerably: on some days a large part of the worm comes forth easily; on other days the part, which you have drawn out or attempted to draw out, clearly resists, and the patient experiences very sharp pains until, as it appears, the subsequent part has been completed and has separated from the part to which it was attached. In this way, the extraction of the worm and the healing of the ulcer require about a month, especially if the body fluids are unhealthy. Hence, three things—entanglement, slow maturation, and extraordinary length—prolong the extraction of the worm. As in the cases of entanglement and maturation, so, too, in the case of length, nothing can be laid down with certainty. I withdrew a worm from the calf of a thirty-year-old man with good results each day, a segment of about a hand's breadth. After three weeks I had not come to the end of the worm. I believe that if one added all the pieces I cut each day the total would be more than twelve feet.

It also happens that a patient has a number of worms developing at the same time. In the course of a single summer I removed no fewer than ten worms from a gluttonous young man who had three times been sent back from Middelburg[11] to the Indies. A number of times I removed three, four, and five worms from a single leg in a period of one month. On occasion, I withdrew two worms simultaneously from one ulcer or two worms successively from one ulcer.

Often a worm comes forth with the middle of its body folded in two while its ends are still concealed inside the patient. If the entire worm matures deep within the sore, the worm putrefies from the prolonged process of ulceration. As a result, about the only vestige of the worm that flows out is the white streaks of corrupt matter which one can observe. It is not always possible to be sure of the number of worms: when a broken worm reenters and a second ulcer is produced at a different location after a number of days along with a considerable excretion of worm segments, one cannot be sure whether this is due to the same or another worm. For when a leg has been corrupted with ulcerous channels and concealed decay, usually a worm is discharged with its parts in reverse order and in a state of putrefaction.

Unfortunate positioning and improper handling sometimes result in a stubborn fistula. On occasion it happens that a worm, after a swelling indicative of its imminent discharge, lies dormant until the following year without inconvenience to the patient; this, of course, is caused by a change of air and way of life. I have treated Armenians, Persians, and Dutchmen who had journeyed on the coast of Hormuz and become infected with the seed of the worm. One year later and having no knowledge of such things, they produced worms in other countries. I knew a man who over a period of two years at Isphahan retained the vestige of a worm, which he had carried from Gamron. It was not before another year that the worm was discharged while the man was returning to Gamron and was only a short distance away. A person who carries the worm-generating virus into a different region experiences much greater difficulty in discharging a worm than a person who discharges a worm in its native area. The nature of the atmosphere in which the worm is born accelerates its excretion and healing, perhaps because the extreme heat promotes putrefaction, which facilitates—quite extraordinarily—the separation of matter from a healthy part. The patient suffers inconvenience only from the afflicted part; otherwise he is healthy, except insofar as the pain impedes his activities.

V
Seminal Principle

I would move directly to the question of a cure, except that a few points need to be made about the origin and seminal principle of the *dracunculus*. That insects and all other kinds of living creatures come from eggs of a similar structural type and that worms living in humans also come from eggs, perhaps those of gnats and flies, is the accepted and firm doctrine of contemporary scientists, who deny that any living being can be generated from an equivocal seed (to use the school terminology). It is difficult to account for the astounding transformation whereby the offspring of a gnat might extend its length in so heterogenous a form. The intestines are not sufficiently narrow as to cause a freshly hatched offspring to develop so thin. And what disposes eggs of the same species so that when hatched in the thick intestines they become ascarides, and when hatched in thin intestines they become lumbrici? One may answer: a more delicate type of egg is hatched more rapidly in the thin intestines, while a more sturdy type of egg from a different little beast is hatched in the thick intestines. I reject this specious conjecture!

It removes one difficulty and replaces it with many difficulties. And although in one way or another you may have surmounted these difficulties with respect to the generation of lumbrici in the intestines, you will still face this question: how can eggs (whether those of earth worms or of other insects) that produce *dracunculi* be protected so as not to be destroyed while circulating the long distance with the blood and be deposited beneath the skin? Posit that this undamaged condition is due to the rapid movement of the humors and the more open ducts in that region. Or if a more restricted path then is to your liking, say rather: invisible particles of eggs (but from what living creature?) fly through the air and enter the very wide pores of the skin, or that as happens with plant buds and tubercles, the particles are carried in by insects too small to be visible. Posit, I say, that all this happens in this way or otherwise in a hot climate. What, then, so vastly elongates a growth from so minute an egg, and what distributes it so widely? Why does the *dracunculus* not come forth full grown in one single act? In fact, however, the initial segment, which is alive, emerges, while the remainder of the worm (just as Pliny's fabled mice are formed from the mud of the Nile)[12] is being created over a period of time. If you fall back on Aristotle's equivocal generation[13] and assert that this crude offspring of nature can somehow be generated with little difficulty from decayed matter or excrement digested three times, or if you assert that it can be generated the way Blanckaert (*Pr. Lib*. 3.10)[14] created a worm for himself according to Cartesian principles, the result is that you will be immeshed in countless difficulties or rather absurdities, which will never allow you to justify your position. This reporter is satisfied to lay the blame on the rain water that is collected from the surface of the land in reservoirs below ground and there stagnates. Such is the annual supply of drinking water furnished by every country where the *dracunculus* is native. I maintain with complete certainty that the seminal principle resides in this rain wash and not in the air or in the food. Europeans who sojourn in this area will contradict my assertion. They drink water that bubbles from fountains and is transported daily over a number of miles. Yet these Europeans contract the worm-producing contagion. The truth of the matter is that often these Europeans are unaware of what they are drinking when they are thirsty and have consumed a considerable amount of brandy or wine; the servants bring any water at hand rather than the water desired. While at Gamron, I learned from personal experience that the camel drivers, instead of bringing to the Dutchmen who were staying there water from a fountain at Isjiin, brought water from a rain puddle half the distance away. I have as supporters of my position the mountain people who inhabit the same region but in the area of bubbling fountains. These mountain people only know the *dracunculus* by name; they never contract it. Very rare exceptions are the individuals who have traveled the desert and imbibed the virus along with the stagnant rain water. I do not, however, deny that the conception of the *dracunculus* is promoted by the nature of the air and improper diet.

VI
Cure

The usual cure employed by the common people is as follows. The induration is ripened with an emollient plaster or ointment in order to gradually lure the worm that is firmly set in the suppuration. The segment grasped—as much as emerged with gentle traction—is wound around a stick about one inch long, and care is taken with an epithem to prevent the worm from receding into the ulcer. The ulcer is cleansed of its pus by changing bandages twice daily. On each occasion, a traction and an attempt at total removal are repeated. Care is taken not to draw too forcefully or suddenly. Improper drawing is futile and causes pain because a worm, which is wound about or fastened to some part, cannot successfully be lured out before it separates itself as a result of maturation. Improper traction is indeed dangerous, because when, as is usual, the worm is broken and reenters the body, it will afflict the patient with many symptoms until it is discharged with a larger suppuration through either the same or a new ulcer. It may happen that one end of the *dracunculus* has been held back by some part of the body other than the ulcer and therefore causes pain from the tension, and over a long period thoroughly resists inducement to emerge because at some distance it is insolubly adhering to an tendinous part. In this event (as I have learned from experience and the discomfort of patients) the more satisfactory procedure is to let go of the other end of the worm and allow it to reenter the ulcer; the entire matter is entrusted to nature. Subsequently, the worm will be extracted with less pain and danger via a new ulcer at the point where it is lodged, or the worm will after a longer period flow out in a state of putrefaction along with an abundance of pus. When you have withdrawn a new segment of the worm, it is permissible to cut off the previously extracted portion so that the size of the rounded mass does not interfere with the bandaging. It is sufficient to retain outside the ulcer as small a part of the worm wound around a stick as is necessary to prevent reentry and to permit another withdrawal at the next bandaging. When the worm has been completely discharged, the fistulous ulcer heals thoroughly and very easily with any ordinary remedy within a few days—quite beyond the expectations of the person conducting the treatment. Impoverished natives manage with good results the whole business of ripening and withdrawal merely by applying onions baked in ashes. Some natives are content to treat themselves by frequent washing with cold water. They claim that this method protects the healthy part from corruption, guards the rest of the body from the discharge, and prevents ulcers from persisting after the worm has been excreted. This rationale, although it may seem crude, is perhaps not absurd for that hot climate. Certainly in Java persons who are suffering from morbilli[15] experience a difficult recovery unless they vigorously bathe with cold water. I know a surgeon at Batavia (the brother of the very distinguished Doctor ten Rhyne)[16] who in one month lost his own three sons whom he was stubbornly treating according to the European method, while a black neighbor saved all his numerous children by employing no remedy but this bathing, which was repeated morning and evening in the open air. My conjecture is that this bathing inhibits the effervescence of harmful matter as well as the emission of fine particles. Since everyone in this land knows that extirpation of the worm is of all importance, it sometimes happens that after an overly hasty incision injury is caused to a tendon that has been grasped instead of a worm; the affected member suffers tragic damage. I have seen two men rendered lame from this crass mistake by surgeons (whose profound ignorance pervades all Asia). Before undertaking the removal of the worm, it is appropriate at the outset of the ailment to evacuate the humors and to moderate them as the ailment progresses, in order to prevent the ulcer from growing worse from the abundance and sharpness of the discharge. In a torrid climate gangrene is easily induced in the shins by the application of fats. It is therefore safest to apply first a cataplasm instead of fat to the swelling and, as the condition progresses, to cover the location with a plaster if only to hold fast the stick around which the end of the worm is wound; no turunda[17] at all is inserted. Our surgeons, as I occasionally remark, thoroughly abuse the turunda. With it they accomplish nothing other than to irritate the part, to excite pain, to rouse evil humors, to impede the outflow of pus, to cause inflammation, and to prolong healing—they are servile and slothful imitators of their ignorant teachers!

Notes

1. The Greek term is *drakon* ("dragon" or "serpent"). The Latin equivalent is *draco*, whose diminutive is *dracunculus* ("small serpent" or "dragonet").

2. Daniel Sennert (1572–1637), physician; see Observation II, note 19.

3. *crinones puerorum*: literally, "boys' hairs."

4. Mitesser: blackhead or grub.

5. Avicenna (980–1037), Persian physician and author of *Canon of Medicine*.

6. Georgius Hieronymus Velschius, *Exercitatio de vena medinensi, ad mentem Ebnsinae, sive de dracunculis veterum* (Augsburg, 1674).

7. Abraham Zacuto Lusitano (1575–1642). Born in Lisbon (hence the Latin "Lusitanus"), he studied medicine in Spain and returned to Portugal to practice medicine for a period of thirty years. As a Jew, Zacuto was forced to flee the religious intolerance of Philip IV and the Inquisition; he settled at Amsterdam where he died in 1642. His works include: *De medicorum principium historia* (Lyon, 1629–42); *Praxis medica admiranda* (Amsterdam, 1634); *Introitus ad praxin et pharmacopaeam* (Amsterdam, 1641); *Epistola de calculo qui gignitur in cavitatibus renum, non in substantia* (Leyden, 1638). His *Opera Omnia* were last published in Lyon, 1694.

8. Jerome Mercuriale (in Latin, "Hieronymus Mercurialis," 1530–1606), *De Morbis cutaneis et omnibus corporis humani excretionibus* (Venice, 1572). A voluminous writer, physician, and professor of medicine at Padua and later at Bologna, Mercuriale was summoned to Vienna to treat Maximilian II. Upon recovering, the emperor richly rewarded Mercuriale with the dignities of knight and count palatine. Mercuriale has been criticized for his acceptance of ancient doctrines and hypotheses as opposed to sound observation, a fault which was perhaps responsible for his denial of the existence of a plague in Venice—an error that countless deaths forced him to acknowledge.

9. Kaempfer notes the division of Persia into five parts, one of which is Chorasmia, or Khorasan, in the east and extending to India (*Amoenitates Exoticae*, 134–35).

10. Paliacatta, almost certainly Palayankottai, a city in Madras, in southeast India.

11. Middelburg: capital of the province of Zeeland in Holland.

12. Pliny notes, "But credibility is given to all these statements by the flooding of the Nile, with a marvel that surpasses them all: this is that, when the river withdraws its covering, water-mice are found with the work of generative water and earth uncompleted—they are already alive in a part of their body, but the most recently formed part of their structure is still of earth" (*Historia Naturalis* 9.84.179).

13. *Abiogenesis* and *generatio aequivoca* are terms for the Aristotelian theory according to which fully formed living organisms may arise from non-living matter.

14. Son of the classical scholar and editor Nicholas Blanckaert, Stephen Blanckaert (1650–1702) was a renowned physician who published voluminously on medicine and surgery. Criticized for indiscriminate borrowing, Blanckaert has left us *Anatomica practica rationalis* (Amsterdam, 1679), which went through many editions and translations. A collection, *Opera medica, theoretica, practica et chirurgica*, was published in Leyden in 1701. Kaempfer's reference is to *Praxeos medicae idea nova* (Amsterdam, 1685).

15. morbilli: measles.

16. Dr. Wilhem ten Rhyne (1649–1700), the celebrated Dutch physician and botanist, made major contributions to science: in 1683 his *De Acupunctura*, the Western world's first detailed treatise on acupuncture, appeared; and in 1687 his classic report on leprosy, *Verhandelingen van de Asiatise melaastsheid*, was printed.

17. turunda: a pellet of some substance used in treating sores.

Plate 2. *Asafetida Plant*

Observation V
Report of Disguun Asafetida

I
Botanical Description

Hingiseh

An Umbellifer related to *Levisticum* and resembling the ramose leaves of the Peony; with a very large and filled caulis; with seed that is foliaceous, bare, solitary, and similar to that of the Bear's Foot or *Pastinaca*; with a root that pours forth asafetida.

The Hingiseh plant has a root: restible for many years; large, heavy, bare; black on the exterior; smooth in lime soil; rough and somewhat wrinkled in sandy soil; usually simple like that of the *Pastinaca*,[1] often branched a bit below the top into two or more divarications, some descending perpendicularly, others extending irregularly and obliquely as if turned by obstacles; the top, rising above ground, is tightly packed, in the fashion of the *Peucedanum*,[2] with fibers rising like bristles, dark red and rigid. Its cortex is rich and juicy, separates easily as the root is withdrawn from the ground, and has a smooth and moist cavity; the substance is heavy, solid, and very white like that of a turnip, full of sap that is rich, very white, very fetid, and strikes the nostrils with a harsh odor resembling that of a leek. The sap collected is called Mingh by the Persians and Indians, Asafetida by the Europeans.[3]

In late autumn, leaves, six, seven, or a larger or smaller number according to the size of the root, spring forth from the top. They thrive luxuriantly during winter and become dry in the fullness of spring. The leaf, however, is ramose, flat, and a cubit long. It very closely resembles the peony leaf in shape and the *Levisticum*[4] in substance, color, and smoothness. Its odor is weaker than that of the sap; its taste is goatlike with bitterness and an aromatic sharpness. Stem and branch comprise its structure. The stem is a span or longer, thinner than a

45

human finger, somewhat striated, fibrous, green, canaliculate around the base from junctures, and otherwise terete. A branch has five, rarely seven, wings, positioned unevenly opposite, rising obliquely, the lower longer than the upper. The wings are divided on either side into several lobes of an uncertain number and unequal size, oblong and somewhat ovate. On some plants the lobes are very narrow, longer, separated up to the costa and by a larger interval, and consequently fewer and solitary, resembling individual leaves; on other plants the lobes are wider, shorter, and continuous due to the rather slight separation. The ends may be incised and are generally ovate or orbiculate, according to the varied inclination of playful nature who often so varies the leaves of individual plants that the leaves scarcely appear to belong to the same species. The lobes grow obliquely, hanging from a thin duct through the sides of the costa, are grayish-green, smooth, sapless, stiff, fragile, and somewhat concave on the front. They have a single thin nerve that runs unevenly lengthwise from the costa and is rarely connected with lateral ones. The size varies but may be three inches long and approximately equal to a thumb in width.

Before the death of the root, which usually occurs in late summer, together with a fascicle of leaves there emerges a caulis. It is simple, straight, terete, somewhat striated, smooth, grass-green, and luxuriantly rising to a height of one, one and one half, or more fathoms. It is thicker at the base than the grasp of a man's fist, gradually becomes thinner from the base up, and separates into a few branches, which in turn separate into umbels as is normal with *Ferula*. The caulis is covered with the rudiments of leaves set alternately at intervals of a palm. Their broad, membranaceous, and turgid bases embrace the caulis unevenly and in decussate fashion. As the rudimentary leaves hang down, they mark the caulis with their imprints, which give the appearance of joints. The caulis is filled with medulla uninterrupted by joints, very abundant, very white, fungous, and intermingled with a few short, stray fibers stretching out lengthwise. The umbels, some supported by a style of one foot, others by a style of a span or less, unfold into quite a few rays (10, 15, or 20), which spread into an orb. The individual rays, as if about to produce subordinate umbels, terminate in a few (possibly 5 or 6) small rays that are two inches long and support bare and upright seeds on short and very thin axils.

The seed is flat, foliaceous, dark red, ovate, not unlike the seed of bear's foot or garden parsnip but a little larger and darker, somehat pilose or rough, marked with three striae, one lengthwise in the middle, the others led in an arcuate course on each side through the fimbria to the tip. The seed has a slight porraceous odor and a heavy and sharply bitter taste. In the middle there is a medulla, or true seed, which is black and compressed with an ovate and mucronate shape. I have not seen flowers, which are said to be very small and pale white; I would suspect they are five-petaled.

II
Extended Report

Avicenna[5] calls the asafetida plant Andsjudaan and Haltiit, which Dioscorides[6] renders as Silphium, and Mattioli,[7] as Laserpitium. In Persia, the plant as well as the gum collected from it is called Hingiseh; and in India, Hiing. It is more usual, however, to call the plant, Hingiseh and the tear, Hiing; these are the terms I use in this description. I will not discuss the origin or development of the term asa. Because of its horrid stench the Germans call it Teufel's Dreck, that is, devil's dung. This plant has been improperly assigned by the conjectures of botanists to various species and elucidated with notes and names by the distinguished authors Scaliger[8] and Saumaise.[9] I here record an authentic report of the plant—the result of my own diligent observation and examination for which purpose I undertook a laborious journey from the city of Gamron[10] to the location of the plant. The plant's only native land is Persia; not Media, Libya, Syria, or the region of Cyrene.[11] Consequently, the distinctions about the sap that authors make on this basis are nonsense. I have heard from two Chinese spice dealers that the plant grows near the walls of China and from it tears are obtained. But because I do not find the plant listed in the Chinese Herbarium, I am uncertain about the veracity of these reports. The introduction of the gum by way of the wall could have generated such an error among the ignorant. Today at least two areas in Persia support the plant: (1) the fields and mountains around Herat,[12] the emporium of the province of Khorasan;[13] and (2) a mountain range in the province of Lar,[14] which extends from the Kur River[15] as far as the city of Kong along the expanse of the Persian Gulf, two parasangs or in other places three or more parasangs from the shore.[16] In both locations, however, a plant that is rich in sap is not found everywhere. The plant found around Herat grows in the flat desert, but the plant found in the area of Lar only grows in the mountains bordering the territory and town of Dusguun, or as commonly pronounced, Disguun. Plants that grow on the near or far side of these mountains lack sap. They are not worth the expense of harvesting, or even if they might be, they are not harvested. Now on the near side of Disguun, neighboring Arabs (colonists from the shores across the sea) have never attempted to harvest the plant. Avoiding any new activities, they are content with the indolent and meager pastoral way of life that they lead beneath their tents. But on the far side, the herb is said to be sweet and to emit only a mild odor so that flocks of goats feed avidly on its leaves and grow wondrously fat on them. To facilitate this process, the animals should be given one feeding of mountain salt (the only kind available here) on the day before they are put to pasture, and then denied any water for the first fourteen days of pasturing. The plant grows promiscuously in wasteland and projections from it, wherever the wind carries the seed torn from the plant; however, the plant grows richer and more frequently at intervals of one foot in places that are somewhat level and more suitable for the seed because they retain and nourish it in soil of higher quality. The plant is seldom found in soil that is moist or fertile with finer earth. More frequently, the soil is sandy, mixed with lime, and is rocky and arid. When surface conditions are inadequate, the root goes deeper to draw water.

The inhabitants of Herat distinguish two species of the plant. The first is called Hingiseh. They assert that it comes from the mountains and woodlands of Disguun. The second, from their own plains, is called Husjeh. The former, they claim, distills a tear that is scant, thin, and rather weak in potency, while the plant from Herat pours out a richer and more fetid tear in greater abundance and one that is

Plate 3. *Asafetida Harvest*

consequently superior. For the sake of truth and to clear up the points of difference, I carefully compared the plant from Herat (which I had obtained in a decayed condition from Khorasan during my stay in Gamron) with the one from Disguun. It would be untrue to say that I observed any difference in shape. I even showed the plant from Disguun, as soon as I could, to the men who carry asa from Herat (which they transport to Gamron almost every year). These carriers, who did not know the source of the plant, affirmed on the spot that it was their own Husjeh, the true mother of the authentic asa. Consequently, I conclude: the plant from Herat does not differ from the one from Disguun except by reason of the soil in which it grows. The soil in the plains of Khorasan is perhaps richer and therefore furnishes its root with a larger capacity for sap. The sterile mountain range of Lar cannot supply its root with such richness. I need not say anything at this point about the tears obtained from each province, which when compared give no hint of a difference of species nor any other difference except those resulting from adulteration[17] or variation in the time or sequence of collection. All this will be clarified rather fully below.

The people of Disguun believe that there are two sexes of the plant: they say the males are devoid of sap, produce a caulis with seed, and as a result die completely; they say females have sap and no caulis. The truth is that this distinction is erroneous and arrived at through carelessness. There is no root found that when wounded before producing a ferula does not pour forth sap. Nor does any root exist that when left to its own fate does not sooner or later rise into a caulis and, having produced a caulis, wither and die from the loss of vital fluid—the process is common to this plant and almost all other Umbelliferae.[18] They say that the root enjoys a wondrous longevity rivaling the human life span; as a result it is not extraordinary for the plant on occasion to achieve monstrous size.[19] They claim that if location permits and the root does not during the first summer grow into a ferula (as sometimes happens), it grows to the length of one fathom and the thickness of a person's wallet; at the midpoint of its life, it has the thickness of the calf of the leg or of an arm by an annual growth rate equal to one thumb; the increase in height always corre-

sponds proportionately. The head, hirsute with fibers, can be an index
of age: I believe that these fibers are the remains of disconnected
leafstalks whose strength even devouring time cannot destroy.

III
Attributes

All asafetida drips from a wounded root; none flows spontaneously
or is extracted artificially from the caulis. Hence, the distinction made
by the renowned Worm into stalk and root asa vanishes.[20] A root less
than four years old is not milky enough and is never cut; the older and
larger it is, the greater will be the abundance of its tears. If cut even on
the day after removal from the earth, it will nevertheless distill milk.
No wonder it overflows with sap, for this is responsible for the
excessive weight. When the root is cut crosswise, the disc-shaped
surface pours out milky sap, and as successive discs are skillfully cut
off, the root continues to produce (see the illustration [Plate 3]). When
carefully observed, the root is found to be composed of two substances: one is firmer with fibers running lengthwise in varying
positions; the other is rather spongy, congeneric, and softer. The latter
substance appears to be designed to retain and prepare the liquid in its
utricles; the former substance is designed to convey and diffuse the
liquid in order to nourish the stalk (and additionally to contribute a
degree of firmness to the brittle root). When the root has dried, all the
softer substances vanish while the fibers remain and contract into a
tow-like pith; the wrinkled cortex, however, loses little of its bulk.

Fresh sap flowing from its utricles is very white, fluid, and rich. It
is exactly like the flower of sweet milk and so is free of any stickiness.
But from contact with air or sun it takes on a darkish color and as it
congeals, develops a stickiness. The stench is an index of excellence;
the stronger the stench, the better the asa. The stench is strongest in
a fresh tear and should in no way be compared with that which is given
off by the hardened and older tear of the type familiar in Europe. I
would dare to assert that one freshly distilled drachma diffuses a greater
stench than a hundred pounds of the older asa which our spice dealers
sell dry. When I returned home (a large and spacious palace with an

impluvium) from the mountain with a quantity of roots, the rooms were so filled with the stench that I was obliged to remove the roots immediately. When a Cafila (the term for a large number of animals carrying wares and accompanied by guides) transporting asa from Khorasan enters the plain, it unloads this cargo far from the gate. The mere unloading, as the wind blows from that direction, causes the air of the entire city to be polluted with the odor. Asa has to be transported to India on a special ship emptied of all wares that are liable to be spoiled by it, since experience has demonstrated that asa taints them and contaminates liquids, especially potables. The ship on which I sailed to Arabia was carrying a single sack of Herat asa hanging from the stern. As the asa penetrated the hide, we were assailed by the offensive odor. The captain feared that on even so short a voyage the rose water, Sjiras wine, and edibles on board would be ruined.

Dioscorides has catalogued a long list of the medicinal values of asa;[21] no less fulsome in praise is Garcia da Orta (*Arom. Hist.* 1.3).[22] Persian physicians, on account of the delicate temperament of that nation, virtually abstain from its use; rustics from Lar at least have, on the advice of the Bengals, learned how to use it for curing gripes, dropsy, and especially tympanites. An inhabitant of Disguun told me that when he was swollen with the last mentioned disease, he consumed every morning for six weeks a lump or a rather large pill of asa until he regained his health. As a result, such stinking odors arose all about him that he was required to avoid the company of men for the entire period. The seed, too, is quite excellent as a medicine but less so than the root. Hence, it, too, is transported from this region by the Indians for medicinal purposes. When fresh sap is applied to wounds, they are said to heal miraculously. When the plant is placed in ditches through which water flows to gardens and palm groves, it kills off little worms of any origin. The Indians, especially the Bengals, make frequent use of asa even for seasoning foods. The distinguished Renodaeus scarcely believed Garcia's report of this; for he says, "If this is not a fable, I conjecture one or the other: either asafetida is not fetid in India or the Indians have throats made of brass" (*Mat. Med.* 1.8.8).[23] I have tasted small pastries dipped in this sap. They did not taste as thoroughly loathsome as I had expected. It is customary for the

Bengals to rub the rim of a cup with this tear in order to stimulate the appetite of a sick person.

There is considerable disagreement between the people of Herat and the people of Disguun about the value of asa. Each of these peoples prefers its own asa and disparages that of the other. The people of Herat say that asa from the mountains of Disguun is thin, weak, and spurious, and they proclaim that their own asa from the plains is much richer, softer, and more fetid. The people from Disguun reply, "Whatever richness Herat asa has is entirely foreign to it." Indeed, the people of Herat mix (to prevent hardening!) with freshly collected asa the flower of camel's milk or of goat's milk, since the asa would not last through its own powers unless sustained by this artificial treatment. And so they shamefully deceive the buyer, who, ignorant of the adulteration, takes the derived richness for authentic richness. Envy and the jealous desire to profit are responsible for this disagreement. It does not merit enough consideration for us to conclude that two different species of plants and tears exist. Consequently, we reject this and the other distinctions made by spice dealers, but we do distinguish by reason of location both all the plants and asafetida of our period into Plain and Mountain, or into Herat and Disguun, or, if one wishes, by name of province into Khorasan and Lar. The former is generally richer, softer, and is shipped packaged in goatskins or sheepskins; the latter is drier and is shipped in sacks woven from wild palm leaves. I will now discuss the collection of Disguun asa as I observed it; the collection of Herat asa is no different except for a few details.[24]

IV
Harvest

This hingiferous[25] (pardon the hybrid term) harvest is undertaken chiefly by the greater part of the citizens of Disguun and also by certain of their rustic neighbors. The town has a population of about three hundred. The harvest is completed in four phases or what amounts to the same thing: four excursions to the range of hingiferous mountains that are situated at distances of from two to four parasangs from the town. I will relate the phases of collection just as they were executed by the townspeople in 1687, the year I visited the mountains. This sequence is never varied, unless perhaps by designating another day for the beginning of the labor or by interposing shorter holidays between collections.

Phase I

Before undertaking the harvest, the collectors inquired as usual whether foreigners were desirous of asa lest there be no reward for their travel, labor, and expenses. When assured of a high price, they flocked to the mountains in the middle of April, since it is about this time that the root is ready to be prepared for pouring forth the tears. The rather gloomy appearance of paling leaves as they turn from vigorous to withering is the sign to begin work. If the neighboring rustics have also decided to harvest, they must be on hand in the same month for this phase. Having journeyed over the vast expanses of the mountains, they separated and stationed themselves at intervals so that those who by prior agreement intended to collect asa for common profit (individual households, related families, groups formed from one or more streets) might occupy one or another tract of the mountain that they assigned themselves for undertaking the harvest.

As the harvest began, each person sprang upon the plants with speed prompted by rivalry. First, armed with a hoe, each worker dug out a circular portion of earth that surrounded the root and that was to be sure compact and sandy or rocky. Excavation to the depth of a palm or a span allowed a small portion of the root to stand exposed. Second, they grasped the leafstalks crosswise and with a twist tore them from the root. In this season the leafstalks come off with no difficulty. They cleared away the fibers on top, which resemble a shaggy turban and interfere with their work. When these are removed, the wrinkled skull is exposed. Third, earth that had been thinned out by hand or with a hoe was replaced around the root up to its top. Over this they placed the leaves that had been broken off and any available type of herb arranged in bundles. A stone was superimposed to prevent the wind, which is often very strong, from blowing the coverings away; it also served as a marker to point out the spot upon their return. Note: This covering is necessary to protect the exposed root from the sun's rays, for once struck by them, the root rots within a natural day and will bring no profit to the collector.

In this manner the citizens prepared tens of thousands of roots for subsequent harvesting. A group of four or five men usually acquired about two thousand roots. Then they left the mountains and returned to their homes; the first phase of their harvesting had been completed within about three days. They call this phase Kusjten, that is, to kill—the killing phase, as it were—because in it the plants are condemned to utter destruction, to be drained in the future of their vital sap until they die.

Phase II

After an interval of forty days (longer this year than was appropriate) the throng of collectors was ready to return to the mountains. They took to the road in the evening and arrived at sunrise on May 25. They withdrew by groups to their own areas to receive their reward of liquid from the roots prepared in the earlier phase. All the liquid, carried upwards to nourish the leaves, now stood stagnant on the top. The collector's instruments were: a very sharp knife for cutting the roots; an iron hand spatula with a wide front for scraping the tear; a bowl or small cup, hanging next to the thigh, to receive immediately the portion removed; and two baskets, carried from the shoulder, for bringing back the harvest of sap.

It should, however, be noted at the outset that each and every group divides the tract of land, which it chose to harvest for itself, and consequently all the roots therein, into two classes. Each class is worked on alternate days while the other is left idle. The reason for this is that when sap has been removed, the root requires an interval of time both for acquiring new liquid to transmit and for a certain necessary thickening of the liquid already emitted. For clarity, the reader should consult the illustration, which shows a tract of the hingiferous mountain large enough for harvesting by a single group. In the illustration [Plate 3] the letter A indicates the first location and class of roots; the letter B indicates the second; on each side is pictured a single root, which, since it is like all the others, may serve as a model. Having noted this, I continue the account.

When the collectors arrived in location A, each collector rushed to a root. The materials that had covered the roots were removed, and with their hands the workers cleared away a certain amount of the surrounding earth, as much as would hinder the operation. When this was done, they sliced off the top of the root, as much as had been hirsute, so that the mutilated root offered a flat disc on which the exuding liquid, to be collected two days later, might stand without danger of flowing off. Next, as in phase one, they again protected the wounded root from the sun, being careful that the covering bundle did not touch the disc. To this end, the bundle must be fashioned into a kind of arch to avoid dispersing the flowing sap by pressing on it. When this task had been performed, on the next day, May 26, at daybreak they proceeded to location B. They cut the roots in the same way as before and again covered them most carefully. On May 27 they returned to location A. They removed the coverings that protected the plants, cleared off the liquid deposited on the surface with spatulas, and immediately placed it in containers that hung at their thighs. Next, they removed a portion of the surrounding earth, as much as would impede cutting, and sliced off the drying surface of the disc. The very thin disc removed is scarcely as thick as an oaten straw. It is sufficient that only the uppermost part of the surface, which closed the pores, be removed in order that the unplugged ducts can deliver sap. Now, they consider it a proven fact that the thinner the slice, the more quickly and more abundantly the sap flows.

Again, they hold a strong conviction about the method of cutting. They insist that the cutting be entirely by repeated strokes and not by a single slash, since the former method promotes a greater flow than the latter. Let the credulous believe it. Note: The collectors immediately empty out the cups hanging by their thighs and place the tears in larger containers or leaves lying on the ground so that the tears may harden the more from exposure to the sun. This also causes the tears, which are by nature white, to take on color according to the susceptibility of the elements and the different locations that unevenly admit the torrid rays of the sun. A fresh lump also takes on very readily an unnatural color from an object placed in contact with it. When the root has been covered over, the task is complete. On May 28 the same work of collecting with all its requirements was accomplished in Location B. On May 29 they returned to location A. The roots were milked for a second time, the earth was removed, and the liquid in location B was collected and a new cutting along with the other necessary operations was accomplished. Now, this labor is proper to phase two, in which each root is wounded three times and stripped of its sap twice. After this has been completed, the entire supply is piled in baskets and carried back with a long pole resting on the shoulder. Each group of four or five men brings back a harvest of ten or twelve Disguun man,[26] which is equal to about fifty Dutch pounds. Note: The tears of the first collection are not considered the best asa, but of inferior quality.

Phase III

After a rest of ten days (eight days were sufficient) had been granted to the roots for regaining sap, the collectors returned to location A at daybreak on June 10 to undertake the harvest again. They removed the coverings and the earth, collected new tears, cut a thin disc from the top of each root and covered the wound. On June 11 the same work on the roots was accomplished in location B. Note: This sap, which after a pause of some days had flowed more abundantly and dried to the required thickness, has its own name, Pispaas, while the other sap bears the ordinary name Sjiir, that is, milky, from its whiteness and thinness. This Pispaas by common agreement is said to be superior and is regarded as more precious than the Sjiir. Whether this is due to scarcity or to a firmer consistency, I could not say. Certainly, Sjiir is not a poorer substance, although it is more moist, for when exposed to the air for a longer time it becomes equally hard and gives no indication of differing from Pispaas. I believe, therefore, that the error arose because the people of Disguun had previously never sold unadulterated Sjiir tears just as they were produced but sold them augmented and adulterated with a cheap material that was easily incorporated into the fresh and fluid tear. The case is otherwise with Pispaas, which due to its dryness rejects adulteration and could only be sold pure. All asa, in fact, is in itself very pure and whatever foreign matter it contains is entirely the result of adulteration. The collectors themselves admitted the following to me. With fresh Sjiir they usually mixed not flour or sagapenum,[27] as has been reported by writers, but pure lime soil, which by chance was available on the mountain. Most collectors added an equal amount of lime, but some added a double or greater amount depending upon their own avarice and how much the thinness of the sap allowed. The detection of this fraudulent practice caused the price to decline and brought utter contempt upon asa of this type. The loss was their own, for no one any longer desired Disguun asa. Subsequently, therefore, having learned to deal more cautiously, the collectors themselves rejected every form of adulteration and today join all the tears of every collection and consistency into one lump. They transport it packed in the same wrappings to the harbors of Kong and Hormuz. If any impurity lurks therein, it is due to the negligence of the collectors who covered the wounded root with insufficient care. Still, it would scarcely be possible to prevent the sticky liquid from being polluted by the leaves that cover it, and by the chaff and the dust blown by the wind. On June 12, class A, and on the following day, June 13, class B, which had exuded Sjiir sap, were cut back and recovered in the same fashion. The same operation was repeated on June 14 in location A and on June 15 in location B, and so, after the roots had bestowed asa, the drier Pispaas once and the more fluid Sjiir twice, they were left beneath their coverings.

Phase IV

On July 3, after three days at home, the collectors returned to the roots, which had not enjoyed a long rest. They had learned from experience that roots, weakened by so many extractions and inclined to perish, waste away easily and will bring no further profit, if collection is delayed too long. On the first day, Pispaas asa was collected, and the other tasks were performed in the usual way in location A. On July 4 exactly the same things were done in location B. On July 5 the collection of Sjiir asa and the remaining tasks were accomplished in location A; on July 6, in location B. At last on July 7 the final touches were given to class A: the sap was scraped away, the cutting was omitted, and finally the uncovered roots were left to die off in a short while from exposure to the open air and the rays of the sun. On July 8 class B experienced the same treatment and was condemned to destruction. With this ended the hingiferous harvest.

Reader, there you have the common method of collection observed to this day in these mountains neighboring Disguun. It comprises three excursions and the extraction from each and every root of Sjiir liquid eight times and of Pispaas, three times. It should be noted that the larger roots, that is, the ones over twenty years old, such as are found in the more distant part of the mountain range, which is approached only with difficulty, are not so quickly abandoned. They can be drained more often until they have poured out Pispaas four or five times and Sjiir, correspondingly more times, since the period of collection extends into September. In this area of the mountains, however, they find few roots older than ten years and none older than twenty. For a good many years during which asa continued to be expensive, the roots were destroyed by collectors and given no opportunity to grow larger. When their tribute has been exacted, the roots, left uncovered, are said to rot without exception. One collector asserted that according to his experience roots buried in the ground sometimes revive; the others denied this.

Notes

1. *Pastinaca*: a genus of apiaceous plants to which the parsnip belongs.

2. *Peucedanum*: a genus of apiaceous herbs to which hog's fennel belongs.

3. *Asafetida* is a Latin word (*asa* = gum resin; *fetida* = bad-smelling). Kaempfer uses the terms *asa* and *asafetida* interchangeably. *Asafetida* appears in English as early as 1398, the year in which John Trevisa completed the translation of Bartholomew de Glanville: "Some stynkynge thynges ben put in medycynes, as . . . Brymstoon and Asafetida" (*De proprietatibus rerum*).

4. *Levisticum*: a monotypic genus of apiaceous plants. *L. levisticum* is the garden lovage.

5. Avicenna (980–1037), Persian physician.

6. Dioscorides (1st century A.D.) was a Roman army physician and herbalist. For *silphion* (Greek for *asafetida*) see Robert T. Gunther, *The Greek Herbal of Dioscorides* (New York, 1959), 328–29.

7. Pietro Andrea Mattioli, *Commentarii in libros sex Pedacii Dioscoridis Anazarbei de medica materia, Opera quae extant omnia* (Frankfurt, 1598).

8. Julius Caesar Scaliger (1484–1558), who made important contributions to botanical, zoological, and literary scholarship, was mistaken as a champion of the Scythian lamb; see Observation I, note 5.

9. Claude Saumaise (1588–1653), a renowned classics scholar, who also learned Arabic for botanical research. Juvenal, Epictetus, and Pliny are a few of the ancient authors to whose works Saumaise devoted his critical labors. Kaempfer here refers to Saumaise's *C. Salmasii Plinianae exercitationes in C. J. Solini Polyhistora* (Utrecht, 1689). Saumaise's *Defensio regia pro Carlo I* (1649) sparked a reply by John Milton, *Pro Populo Anglicano Defensio* (1651).

10. Gamron (variously spelled) was renamed Bandar Abbas ("The Port of Abbas") by the Shah Abbas, and replaced Hormuz as the chief port of his realm in the seventeenth century. For the activities of the Dutch in this area see Sir Arnold T. Wilson, *The Persian Gulf* (London, 1928), 153–70.

11. References to asafetida from Cyrene occur in a number of classical Latin works. Perhaps the two most interesting sources are Pliny the Elder's *Historia Naturalis* and the seventh poem by Catullus (84–54 B.C.). Pliny's account furnishes a mixture of fact and fiction: (1) the plant first began to grow in 611 B. C. in the Gardens of the Hesperides and the Greater Syrtis after a sudden shower of rain the color of pitch; (2) the remarkably important plant was

commended for general and medical use; (3) it was worth its weight in silver denarii; (4) during his dictatorship, Julius Caesar found 1,500 pounds of asafetida in the treasury along with gold and silver (19.15.38-45). Catullus' seventh poem suggests the exotic quality of asafetida: You ask me, Lesbia, how many kisses of yours are enough and more than enough for me. As many as the great number of Libyan sands lying in asafetida-bearing Cyrene . . . or as many as the stars which, in the silence of night, look down on the furtive loves of men: to kiss you so many kisses is enough and more than enough for your love-mad Catullus" (1-10).

12. No longer of commercial importance and now part of Afghanistan, Herat is famous for the gigantic earth work upon which the city wall is built. The earth work is 250 feet wide at the base and 50 feet high. The wall is 25 feet high and 14 feet thick with 150 semi-circular towers.

13. The province of Khorasan lies to the southeast of the Caspian Sea.

14. The city of Lar is located at an elevation of 3,000 feet on the caravan route to the port city of Bandar Abbas, formerly Gamron.

15. The Kur River in the area of Persepolis.

16. Kong, located to the west of Bandar Abbas on the Persian Gulf, was the chief Portuguese settlement in Persia. A *parasang* is an ancient Persian measure of length equal to about 3.5 miles.

17. *The Dispensatory of the United States of America* notes: "Asafetida is often purposely adulterated, it frequently comes of inferior quality, and mixed with various impurities, such as sand, stones, galbanum, ammoniac, gums, gypsum, vegetable tissues, or a rose-colored marble. It is generally conceded to be the worst adulterated drug upon the market" (24th ed. [Philadelphia, London, Montreal, 1947], 113).

18. Umbelliferae (or Ammiaceae), a family of plants comprising the carrot, celery, parsnip, parsley, and other similar plants.

19. The Greek poet Hesiod (c. 700 B.C.) reports that Prometheus stole fire from heaven and carried it to earth in a giant fennel stalk (*narthex*), a ferula related to Kaempfer's plant (*Theogony* 565–67).

20. Ole Worm (Olaus Wormius, 1588–1654) was a Danish anatomist after whom the Wormian bone, collectively, small irregular bone structures found along the sutures of the cranium, is named. He held professorships of Greek and medicine at Copenhagen. For his collection of rarities see *Musaeum Wormianum* (Leiden, 1665), edited by his son William, one of Worm's sixteen children from three wives.

21. Among the ailments Dioscorides mentions are: toothache, bites of dogs

and scorpions, wounds caused by poison arrows, carbuncles, corns, convulsions and ruptures (*Materia medica* 3.94).

22. Garcia da Orta, *Coloquios dos simples, e drogas he cousas medicinais da India* (Goa, 1563). Clusius abridged and translated the work into Latin, *Aromatum et simplicium aliquot Medicamentorum apud Indos nascentium historia* (Antwerp, 1567). See *Colloquies on the Simples & Drugs of India, by Garcia da Orta*, new ed. (Lisbon, 1895) ed. Conde de Ficalho, trans. Sir Clements Markham (London, 1913).

23. Joannes Renodaeus, (Jean de Renou, fl. 1608) *Dispensatorium Galenochymicum continens primo J. Renodaei institutionum pharmaceuticarum lib. V., de materia medica lib. III., et antidotarium . . . Secundo J. Quercetani pharmacopoeam dogmaticorum restitutam. Per P. Uffenbachium . . . revisum et . . . medicamentorum ac experimentorum descriptionibus, ex manu scriptis . . . doctorum virorum locupletatum* (Hanover, 1631); *A Medicinal Dispensatory, containing the whole body of physick; discovering the natures . . . of vegetables, minerals and animals, the manner of compounding medicaments: methodically digested in five books of philosophical and pharmaceutical Institutions; three books of physical materials . . . Together with a Pharmacopoea, now Englished and revised by R. Tomlinson* (London, 1657).

24. With these words Brigade-Surgeon J. E. T. Aitchison issued his summons in 1890 to further research on the harvesting of asafetida: "Dr. Bellew's and my own report combined give a very fair general idea of how the drug is collected, and to what other uses the plant is put; but it would be most interesting to see the whole stages gone through, and this could only be done by residing the entire season at one of these Asafaetida-producing districts, along with some of the great nomad tribes, who make the collection of this drug one of their sources of livelihood" ("Notes to Assist a Further Knowledge of the Products of Western Afghanistan and of North-Eastern Persia," *Transactions of the Botanical Society, Edinburgh* 18 [1891]: 72). The same study was also published separately (Edinburgh, 1890). Aitchison served on the Afghan delimitation commission of 1884–85. He believed that the honor of the first published report of the asafetida harvest belonged to Dr. Henry Walter Bellew, C.S.I. The account that Dr. (later Surgeon-General) Bellew compiled when attached to a mission to Afghanistan in 1857 furnished a useful but very brief description of a portion of the harvesting. Dr. Bellew's account appeared in R. H. Davies's *Report on the Trade and Resources of the Countries on the North-Western Boundary of British India* (Lahore, 1862). Appendices vii and xii are quoted in Aitchison, 71–72. For Bellew's journey see his *Journal of a Political Mission to Afghanistan in 1857, under Major (now Colonel) Lumsden; with an Account of the Country and People* (London, 1862). Aitchison's own report is again a helpful but short notice based only partially on firsthand observation. To both

we are indebted for nineteenth-century contributions on the collection of asafetida on the high plains of Afghanistan. But it is to Kaempfer that we are indebted for the detailed first account of the entire harvesting process.

25. hingiferous: hingiseh-bearing.

26. man: a maund, a greatly varying unit of weight in India, Persia, Turkey, and elsewhere.

27. sagapenum: a gum-resin obtained from a species of *Ferula*.

Observation VI
Dsjerenang, or Dragon's Blood, Obtained from the Fruit of the *Palma Conifera Spinosa*

I
Account of the Sap

It is a matter of certainty rather than probability that just as stones resembling bezoars are obtained from animals of different species (the ape, the goat, and the stag) so also are certain resins and gums obtained from plants of different species.[1] While I will here not discuss turpentine, bdellium,[2] tragacanth,[3] and natural varnish, I will demonstrate my point solely in the case of a sap called dragon's blood.[4] Pharmacists furnish us with two types of dragon's blood.[5] The difference lies not merely in purity, as is believed, but in the types of trees that are the sources. The first type of dragon's blood is imported in lumps from the Orient as well as from America. The second type is imported in drops (so we call them) only from the Orient. If we are to trust the accuracy of de l'Escluse, the first type of dragon's blood flows from the openings in the bark of a certain plum tree, called the Draco, which grows in the Canary Islands.[6] If we are to believe the Arabs and Persians, the same substance is obtained from the red sandalwood tree. The Arabs and Persians hold this belief because of the following elements that are common to the tear drops obtained from the red sandalwood tree and those obtained from the sandalwood tree: color, quality, potency, and last the countries of origins—Cholomandel, commonly called Cormandel (Coromandel), Madagascar, and Ethiopia, or rather the coastal lands of Ethiopia called Sangibar or Zanguebar (Zanzibar). The close agreement of the terminology employed by various nations supports this position: the term that the Persians employ for dragon's sap is Chuni Sengijoon, which means Zanzibar blood; likewise the natives of the Ganges have the following terms for the wood of the red sandalwood: Reket Tsjandan (Blood Sandalwood) in the common language of the Deccan; Rakta Tsjandoonam (Blood-wood) in the educated Sanskrit of the Brahmans; or Rakta Tsjan-doonam Sangghi (Zanzibar or Ethiopian Blood Sandalwood). Others also maintain that dragon's blood sap is derived from different trees. This disagreement about the source of dragon's blood is not surprising. Uninformed investigators have ample opportunity for speculation and error because there are many forms of adulteration and indeed a class of substitutes, going under the name of dragon's blood, which are artificial compounds sold for encaustic and painting.

I have done research on the second type of dragon's blood, called blood in drops, in the countries of the dark Orientals, the Malayans and Japanese. I discovered that dragon's blood is artifically extracted from a tiny, strobile-shaped fruit of a very spinose, single-stemmed tree. The technique of extraction is as follows. The fruit is placed on a grid that rests above the water that half fills a large earthen jar. Next, the jar is partially covered and placed over the burning coals. As the steam rises from the boiling water, it softens and shrinks the fruit. In the process, the bloody dye, which was not visible to me in the cut fruit, is drawn by the hot vapors, flows out, and stands on the surface of the fruit. After the dye has been scraped off with a stick, it is stored in follicles of folded arundinaceous leaves. Soon afterwards follicles are tied with cord in a series and hung in the open air until the blood within has thoroughly dried. Others accomplish the task simply by boiling. The strobiles are boiled until all the red comes out into the water. Then the strobiles are removed, and by additional boiling the liquid is reduced to a thick sap. The sap is wrapped in follicles as above.

II
Botanical Description

Before describing this tree, I feel it necessary to make the following preliminary remarks. The Indies produce many species of reeds: reeds that have a filled stem are called Rotang or Rottanj [rattan]; reeds that have a hollow stem are called Bambuu [bamboo]. To both groups belong various wild palms that grow in the vast jungles (the haunts of tigers and inaccessible to humans). Upon undertaking a botanical excursion with some natives,[7] I encountered chiefly three rather impressive species armed with spines: Rotanj Salag, Rotanj Gelag, and Rotanj Dsjerenang. Unless my guide was mistaken and confused the names of the three trees, I here record an accurate but very hasty description (the best the skill of this traveler could provide) of the last tree, the Rotanj Dsjerenang, which is important because it yields dragon's blood.

Rottanj Dsjerenang is a palm-pine at most three fathoms high and bristling on all sides with dark brown spines that are straight, nearly one inch long, flattened, and thin. The stem climbs to a height of three fathoms, is as thick as a man's arm, simple, straight, and pale yellow. It is armed with horizontal prickles most dense at the lowest part of the stem and is jointed at intervals of a span, but the joints are not visible because of the bases of the branches that envelop them. The bases of the branches are tubular, each base growing from a joint in such a fashion that the lower base always enfolds the starting point of the one above. As a result, the joints do not show unless the coverings are removed. These coverings, or tubes, constitute both the outer surface of the stem and its greater part, for when the coverings have been removed, one can see the pithy innermost part.

This innermost part has a shiny surface, a deep reddish brown color, and a loose and fibrous substance that at the top is more solidly carnous, edible, tasteless, and very white. In view of the number of joints, branches on the stem are rare and show no pattern. The top has more branches, but they are not numerous; the exterior ones are, as is typical of palms, longer, while the interior branches are shorter, not fully grown, and represent successive stages of growth. The branches are six feet long and adorned on both sides with a pinnate arrangement of leaves. However, the lowest portion of the branches, called the scapus, is bare. The costa of the branches is light; initially thick, it gradually becomes thin and is somewhat flattened. Its top surface is grass-green, while its bottom surface is pale yellow. The costa is carinate on both sides where leaves grow. It is armed with spines that are short, rare, and curved and paired on the dorsum like horns.

The leaves, which botanists commonly call feathers when they use the term leaves for branches, are arundinaceous, grass-green, one cubit long, and one-half inch wide, extended in a long tip, pendulous, and rough on the underside with sparse prickles. The leaves have three nerves running lengthwise: a prominent one in the middle, and on either side a more delicate one, whence just as many turn in folds, and as they wither, somehow they unite. Nature bestows special care on the production of the fruits: the fruits are found in the coma on racemes that grow from the axils of the branches, while on a palm stem, the racemes grow at a distance from the branches. The racemes are loricated with facing follicles that are leafy, thin, striated, dark, and sharply lanceolate in a long tip. The lower and more external follicle exceeds two spans in length and one and one half inches in width, and on the convex portion is rough with one-inch spines that are flattened and reach out at a right angle. The other follicle, which is the upper one or the one facing the stem, is bare and shorter. The raceme itself is nine inches long and is composed of several (four, five, or six) smaller racemes, which closely surround the middle style lengthwise. Each of these smaller racemes is separated by its own follicle. Gradually shorter and narrower follicles (like those described above) separate individual racemes from each other in such a way that subsequent follicles are successively protected by the embrace of earlier follicles. The racemes divaricate in small rods or pedicles that are short, thick, firm, bent back to alternate sides at very short intervals, and remarkable for their tubercles. The tubercles are slight, scaly, and nondeciduous. On the tubercles stand the fruit, which have six laciniae (the remains of the perianth) attached beneath the base. The laciniae are thin, membranous, and dark; the exterior three laciniae are very short, broad, and dull; the remaining three laciniae, which project a little from the spaces between the exterior ones, are longer, narrower and acuminate.

The fruit itself is oviform, larger than a filbert and loricated from the top down with very shiny little scales so that it resembles an inverted strobile; the upper mucrones continuously cover the spaces between the lower ones with the studied beauty of a quincunx. The top of the fruit is adorned with three styles, which are thin, dried, somewhat hard, and elegantly bent back in circles. The little scales are tiny, somewhat hard, thin, tightly adnate, purple-red, dark at the edges, and mucronate at a right angle. When the scales have been removed, one can see a carnous and white membrane. This membrane loosely encloses a carnous globule, which in an unripe state is pale green, pulpy, juicy, and leguminous as well as highly astringent in taste. At the touch of the tongue, the taste spreads through the gums and walls of the mouth. This effect, however, passes very quickly.

After much labor, de Bondt printed a partial illustration of this raceme under the Malay generic term, Rotang.[8] Bauhin furnishes a quite similar illustration under the title Palmapinus.[9] Both men are unaware that the tree produces dragon's blood. The last book of *Hortus Indicus Malabaricus* also illustrates similar scaly fruit but they are from a smaller reed that lacks spines.[10] I have in my collection two fruits that the ocean deposited upon the shore of Amboina.[11] One can hear the sound of the globule enclosed in each fruit. Some people think these are products of the sea; I think they are the genuine fruits of the tree I described and that they were petrified in the sea.

Notes

1. Bezoar is a concretion found chiefly in the alimentary canal of certain ruminants. It was supposed to be of significant medicinal value, especially as an antidote to poison. Kaempfer devotes considerable attention to bezoar in *Amoenitates Exoticae* 2. 9, 381 ff., with illustrations of stones and animals.

2. bdellium: a fragrant gum-resin resembling myrrh. It is obtained from certain burseraceous plants.

3. tragacanth: a gum obtained from various Asian shrubs of the genus *Astragalus*.

4. The substance is in reality an exudation from a plant. Pliny's explanation of the term *Indian Cinnibar* (dragon's blood) makes clear the erroneous origin of the term: "Now this is the name the Greeks give to the gore of a snake crushed by the weight of dying elephants, when the blood of each animal gets mixed together" *(Historia Naturalis* 33.38.116).

5. The following sources of distinct varieties of the drug were compiled by Torald Sollmann, "A Sketch of the Medical History of Dragon's Blood," *Journal of The American Pharmaceutical Association* 9 (1920): 141:

1. *East Indian Dragon's Blood*—This is the ordinary variety of commerce; a resinous exudation from the surface of the fruit of a number of small palms of the genus *Daemonorops*, formerly *Calamus*; growing in the East Indies, Malays, Sumatra and Borneo. The species differ with the localities. It occurs in commerce as "tears, of the size of a hazelnut to that of a walnut; as the familiar "sticks" or "reeds"; and as large "lumps."

2. *Socotran or Zanzibar Dragon's Blood*—Derived from *Dracaena Cinnabari or schizantha*, a large tree of Somaliland. Occurs as tears. Only small quantities reach the markets.

3. *Canary Dragon's Blood*—Derived from *Dracaena Draco*, a liliaceous tree resembling the Yucca. Gathered from incisions of the trunk. Not found in commerce.

4. *West Indian Dragon's Blood*—Exudation from *Pterocarpus Draco* (Papilionaceae?).

5. *Mexican Dragon's Blood*—From *Croton Draco*.

6. *Venezuelan Dragon's Blood*—From *Croton gossypifolium.*

6. Charles de l'Escluse, *Exoticorum libri decem* (Antwerp, 1605), 330–31. The tree in question is the *Dracaena draco*, the dragon tree of the Canary

Islands. The dragon tree grows to a height of sixty or more feet and is famous for its longevity. One tree, blown down in 1868, was believed to be the oldest tree in the world. It was 70 feet high and had a girth of nearly 45 feet. Small greenish flowers are followed by orange berries. The dried sap, dragon's blood, was employed for coloring varnishes and for medicines.

7. For a broad survey with a bibliography of early botanical exploration, see William T. Stearn, "Botanical Exploration to the Time of Linnaeus," *Proceedings of the Linnean Society of London* 169 (1958): 173–96. Stearn rightly says of these botanists, "As men they were as varied as human nature, learned like Kaempfer, gentle like Rumpf, quarrelsome like Samuel Browne, and so on, but they all had perseverance, enthusiasm and stamina. They needed all three" (180).

8. Jakob de Bondt, *De medicina Indorum lib. IV* (Leiden, 1642). This work includes the first description of beriberi.

9. Jean Bauhin, *Historia plantarum universalis nova et absolutissima* (Yverdon, 1650–51), 399.

10. Hendrik Adrian van Reede tot Draakestein, *Hortus Indicus Malabaricus,* 12 vols. (Amsterdam, 1678–1703); see vol. 12, pp. 121 and 123, illustrations 64 and 65.

11. Amboina was named by the Portuguese who founded a settlement there (1521), but the Portuguese policy of suppression made it possible for the Dutch to gain control. The clash of British and Dutch trade interests resulted in massacre by the Dutch on which John Dryden composed a tragedy, *Amboyna, or the Cruelties of the Dutch to the English Merchants* (1673). A monument on Amboina commemorates the naturalist Rumphius (G. E. Rumpf, 1627–1702, author of *Herbarium Amboinense*) who worked and died there.

Observation VII
Andrum, or Hydrocele
(Tumor of the Scrotum):
A Disease Endemic and Common
in the Region of Malabar

It is an obvious fact that water, due to the depths or soil in which it originates, generates diseases that are called endemic, or native.[1] The springs of the Ukraine are responsible for Polish plait.[2] The waters of Styria[3] and Transylvania spread struma.[4] The Arabs of the desert, the inhabitants around the Persian Gulf, and the natives of African Guinea contract *dracunculi*[5] from stagnant rain water. But natives who drink the waters from the depths of Malabar receive two types of disease. The first disease lodges in the scrotum; the second, in either foot. The natives call the first disease Andrum; our people call it oscheohydrocele. The people of Malabar term the second disease Pericàl. It is a type of pedarthrocaces or spina ventosa. I will discuss each related disease in its own chapter in order to avoid the tedium of one long and tiring report.

I
Report of the Disease

Andrum or, as paraphrased, Andu waja ku, means endemic hydrocele. This tragic disease begins with the reddening of the skin of the scrotum that recurs with the new moon of each month.[6] When the reddening ceases after one or two days and the lymphatics have been eroded, a certain liquid gradually flows into the cavity of the scrotum. The fluid, which is serous, a bit salty, and increases daily especially as the moon grows larger, so distends the scrotum that finally the fluid must be drawn out by lancing. The fluid is clear and thin, sometimes rather viscous, often reddish, or diluted with gore, and additionally stained with other elements, according to the constitution of the

patient and his lifestyle. The natives claim that the cause of the disease
lurks in the water that is furnished by the realm of Cocin (commonly
spelled Cochin and pronounced Cotsjiin)[7] from its own wells and
springs which bubble with a muriatic and corrosive salt that is not
found in the other waters of Malabar. After being supplied by the
water, the cause of the disease is nourished by the local air which is
polluted by the unwholesome vapors from the wet depression of the
land. The nearby range of mountains prevents the vapors from
dissipating.[8] These vapors in conjunction with the human body so
dispose and actuate the matter that has been abundantly supplied that
when the vessels of the scrotum, rather than those of another organ,
have been eroded, hydrocele and not some other species of hydrops
can be produced. The wind from the mountains is, in my opinion, an
important factor. While the sun remains beyond the equator, the
wind, rising after midnight, spreads a slight chill through the air and
suddenly closes the pores of an exposed human body, which in this
region are normally wide open. As this wind (for I have experienced
its peculiar injuriousness) thoroughly penetrates rooms, it aggravates
diseases and causes serious convulsions that in Europe would be
thought to herald imminent death but that in Malabar are not at all
fatal. The disease is equally common to natives and visitors who have
spent some years here.

Normally in this location the disease is incurable and lasts the entire
duration of a man's life. However, the disease is not at all dangerous,
nor is it especially troublesome for those accustomed to it, except
perhaps that when inveterate it may (and this often happens) injure
the testicles and even degenerate into hydrosarcocele. As the climate
in which the disease originated alters, the water that accumulates each
month immediately slackens its flow into the scrotum and in time
completely disappears. But the sarcocele already contracted is removed
with very great difficulty or is not removed at all. The virus of each
disease, they say, can be destroyed and obliterated at the source, if the
wells are filled with sand that must be taken from a river in the region
called Mangatti. The river's course flows down from the mountains,
carrying harmless and very clear water. This technique is employed
with some small success by certain of the more cultivated citizens of

the city and realm of Cochin. In Malabar, however, there are two renowned regions called Mangatti. The first Mangatti is named after its important river and is located on the northern border of the territory of Cochin where one crosses to Craganore, the heavily fortified Dutch garrison on the seacoast. The second Mangatti is situated to the east on the inland boundary of the realm of Signatia.[9]

The second Mangatti is exceedingly famous for its sanctuary, or inn, of apes. As an act of piety the Brahmans have trained apes to come from the forests and gather there every day for meals. Rarely does the inquisitive traveler (pardon the digression) have an opportunity to witness the apes fighting, a show the curators of the institution reserve as a reward for individuals who made an offering for maintenance of the beasts. No wonder. When it is mealtime, a staff and cooked rice are set out. The first ape to arrive snatches up the staff in order to ward off the others from the platter of food. Then a brawl arises as some apes grab the weapon and others the food. You could not ask for anything more amusing. The uneducated natives believe that apes are a race of human beings who inhabit the forests but are unwilling to speak lest they be captured and enslaved to people. As evidence, they point out some remains of a crocodile, which hang from a tree by the river's edge. They say that this crocodile was once captured and hung up by the apes with more than human cunning. Now, this predator had devoured an ape who was the one from his group who was nearest to the shore and was perhaps drinking from the river. The other apes, well aware that the crocodiles habitually returned to that spot in search of prey, wove a noose from bark, and strung it from a tall tree down into the water. Such was their trap. Soon, these glorious avengers of their comrade caught their enemy with a noose around his neck, pulled up the rope, and hung him.

II
Cure

Insofar as a cure is concerned, the natives employ a palliative rather than a true remedy. The traditional procedure is to draw off with a lancet the serous liquid from the swollen scrotum either each month or every few months. During the operation, the one and only caution observed is that the blade of the lancet be inserted into the lowest point on either side of the scrotum. There is no doubt that by strictly observing a diet prescribed by a physician and by the faithful use of drugs, which in one way or another promote the desired end, the serum diminishes and collects slowly in the scrotum. But whatever I and others tried with mercury, we never observed eradication of the disease in that region. Since the natives have so often been frustrated in attempting a cure, they prefer to employ the tip of a scalpel, which involves no inconvenience or danger, rather than to obey the many dietary restrictions and physicians' prescriptions, whose results are uncertain and which may even aggravate the disease. Nevertheless, the pagan physicians attempt a cure with epithems of the local types. The following recipes, which I here record in Latin, were said to be representative of all the others and to possess outstanding powers. They were generously furnished to me by a certain physician, a gymnosophist. Would that he were not a sophist!

Recipe. Fry over a fire aaq.s. of oil expressed from sesame, oil expressed from castor-oil seed, called (incorrectly) *oleum cicinum* by de l'Escluse,[10] and melted butter. With the fry anoint leaves of the castor-oil plant previously separated from the pedicles and deveined. Apply these to the scrotum as hot as the patient can endure them.

The following is considered more effective and very potent against the ailment especially when it is chronic:

Recipe. Q.s. of dragon's blood; aaq.s. of coriander seed, and of the bark of the moringa (called *Arbor Alexipharmaca A. Cost.*),[11] individually pulverized in advance on a stone, are to be kneaded into a mass with lemon juice. Apply this with a linen cloth to the scrotum and remove it whenever it begins to burn.

NOTES

1. Endemic hydrocele is a manifestation of mosquito-borne filariasis usually caused by *Wucheria Bancrofti*. Hydrocele is a frequent accompaniment of filariasis because of the common localization of the adult worms in the epididymis. In a smaller percent of the infected persons a true elephantiasis of the scrotum may develop gradually after repeated attacks.

2. Polish plait: a matted condition of the hair due to disease. According to Blancard's *Physician's Dictionary*, 2nd ed. (1693), it is "an epidemical disease in Polonia, when Hairs grow together like a Cow's Tail."

3. Styria: a crownland and duchy of the Austrian empire.

4. struma (scrofula, also called King's Evil): a form of tuberculosis characterized mainly by chronic enlargement and degeneration of the lymphatic glands.

5. The *dracunculus* is the subject of Observation IV.

6. The only periodicity relating to filariasis is nocturnal.

7. The state of Cochin was part of the ancient Kerala kingdom on the Malabar coast. In 1502 Vasco de Gama founded a factory in the town of Cochin where in 1503 the Portuguese constructed the first European fort in India. With the capture of Cochin in 1663, the Dutch put an end to Portuguese dominion and retained control for over a century.

8. The Western Ghats, fronting on the Indian Ocean, rise up abruptly to an average height of 3,000 feet.

9. Craganore (Cranganore, Kranganur, Kodungalur, ancient Muziris) served as a port for Cochin at the mouth of the Periyar river. In 1523 the Portuguese built a fort at Craganore, which the Dutch captured in 1661. The Periyar is probably the river here mentioned by Kaempfer, but the two Mangatti and Signatia could not be identified.

10. Charles de l'Escluse; see Observation VI, note 6.

11. moringa (*Moringa pterygosperma*): the ben nut tree (horseradish tree), native to India. Oil of ben is obtained from the winged seeds (ben nuts); the taste of the root resembles that of horseradish.

Observation VIII
Pericàl, or
Ulcerous Hypersarcosis of the Feet,
or Pedarthrocaces,
an Indigenous Disease of Malabar

I
Report of the Disease

The second disease caused by the waters of Malabar is an edematous tumor which attacks no other part of the body but the foot.[1] The increased monthly swelling so enlarges the foot that finally it resembles the shape of an elephant's leg. The natives call the disease Pericàl or, when the word is spoken quickly, Pircàl, which means feverish foot, and should be understood as a monthly erysipelas of the affected foot. Brahman physicians catalog it as the eighteenth species of fevers. The malady frequently burdens St. Thomas Christians, so named after the apostle.[2] Because they embraced the doctrines of St. Thomas, they were treated with such cruelty by the pagans that they left their own country, Carnatik (in the region of Coromandel),[3] crossed the range of mountains that forms the border, and arrived in Malabar.

They now have more than fourteen hundred villages and as many churches in Malabar. Popular tradition embodies the belief that they carried the disease with them from the land of the Ganges. This tradition is absolute nonsense, unless one believes that fountains and rivers, the sources of the endemic disease, migrated along with the Christians. The natives tell the following story. When the pagans were unable to move a huge tree that had blocked the harbor, St. Thomas dragged it away with his loincloth. Thereupon, those who had converted from paganism to the apostle's doctrines were marked with an elephantine foot by the pagan gods.

This tale so captivated the belief of foreigners that the Portuguese in India readily term the disease Peju de Santo Thoma, and the Dutch, following their example, term the disease Sünt Tomas Been. I found this same tumor endemic (although not as prevalent) in the villages of Ceylon beyond Galle, the heavily fortified Dutch city,[4] as well as in Omura, a region in the kingdom of Japan.[5] In both places the swelling is bred, as the inhabitants admit, by the bad water. The disease never attacks any limb but the leg, one leg or the other (rarely both) and especially its lowest portion. Each month as the moon grows, the disease inflames the foot. Within several days the inflammation cools down, but the swelling from the inflammation does not remit. Instead, it makes its way into the corrupt flesh and causes the circumference of the leg to enlarge as time passes into a deformed thickness that exceeds the normal shape (two, three, or more times). This thickness that I am discussing is uneven, edematous, somewhat hard, and scirrhous in appearance; in places, and this depends upon how the bone is affected, it is fungous, ulcerous, and bubbling with serous fluid. Often the swelling extends through the front of the foot to the very toes themselves; it rarely ascends above the ankles and never reaches the knees. The sources of water, which are located not so much in the whole of Malabar as in the realm of Cochin,[6] furnish the causative agent of the disease. This water abounds in corrosive salt (the local term is natron) whose sharpness, coupled with viscid blood, obstructs and corrupts the limb it ought to nourish. This tragedy, however, develops in the legs rather than in other places because the diseased material by its own weight floods the legs, unless it is intercepted by the cavity of the scrotum when the lymphatics have been eroded. In fact, all persons who suffer from hydrocele are immune to this malady. It attacks chiefly the Christians because as a rule they live around the mountains, whose air and water more effectively promote swelling.

I have no doubt that as part of this condition or as a consequence of it a process of decay, which baffles all medical zeal and labor, has contracted the adjacent bone. It is uncertain whether the bone is afflicted with caries or only an imbalance of fluid. Again, if caries is present, is it common, by which I mean, is it induced from without after the periosteum has been damaged? Or, in fact, is the caries generated from within by a stagnant, viscid, sharp, and acid fluid that nourishes it in the tiny ducts of the bones? Now, this sort of fluid perforates and erodes the substance of the bone. It induces the configuration that the ancients because of the unevenness and sponginess called spina ventosa and ventositas spinae. Severino castigated the barbarity of this name in his work, *On the Hidden Nature of Abscesses*, and substituted the term pedarthrocaces, which means a child's abscess, because the condition is rather frequent among children due to softer bones and their clearer little ducts, which carry a larger supply of nutrition for growth.[7]

Briefly, my position is as follows. I have not operated on a foot, nor would I have dared to determine the state of an afflicted bone. Nevertheless, I am led to believe that the condition should be considered a kind of pedarthrocaces since it most especially attacks younger people and spares older foreigners whose legs are firm and require no nutrition for growing bones. Although the malady is hard to the touch, blackish to look on, and loathsome as well, it does not easily degenerate into gangrene, and indeed involves no danger. The disease also is virtually free from the discomforts of pain apart from that brought on by the monthly inflammation. Nevertheless, the monstrous mass of the malady is hateful and most annoying to persons burdened with it, especially if the small ulcers are wet and dripping as is usual in inveterate cases. You would be amazed at how little it impedes workers on the job. They have become so accustomed to bearing this burden that they are no less suited to performing all sorts of work for hire, and indeed they climb the tallest nuciferous palms with the same agility as do other men.

II
Cure

Patients burdened with this monstrous ailment have no means of relief in this area other than that which is sought after through chronic sinuses. These sinuses must be burned into the flesh about the knee of the affected foot and at the onset of the disease after one or a few bouts of inflammation. Hence, you will frequently observe Christians of this region bandaged with materials to promote drainage. The sinuses do not extirpate the disease that has taken root, but in order to restrain the wanton growth of the swelling and ulcers, the flow of the serum is somewhat checked and intercepted, as if a cataract has been set up. The native physicians, however, attempt a complete cure at least at the onset of the disease; of course, they fail to obtain it. Their method is first of all to let blood, next to anoint the head frequently with oils (coconut oil, common in this land, is preferred to others), and then to purge the body adequately with a drug. When these virtually routine preliminaries have been completed, they surround themselves with water and fire for healing the affected foot. With this double sacrament they hope to regenerate that part of the patient. I here record the ritual that a physician, a St. Thomas Christian, disclosed to me as he read it aloud from his Sanskrit manual, though, by heaven, it strikes me as more worthy of laughter than of approval.

Prepare a small torch made from tow and coconut oil, which the land of Malabar supplies in abundance. Also have in readiness a liquid obtained from ground Corveile leaves together with sweet water (the plant is locally very famous for its large edible fruit).[8] When these have been readied, elevate the foot and then strike measured blows on the sole of the foot with the flaming end of the burning torch, and after each blow sprinkle the sole with a spoonful of the aforementioned liquid. The blow and sprinkling must be continuously alternated for a period of three hours.

If after this exorcism has been performed once or twice, your guest disease is not driven out, whatever further effort you make to expel him, said the physician, is in vain. The Japanese of the province of Omura, when attacked by this disease (called Kojejas; Kojassi by the common people), liberally scarify the initial swelling which then usually subsides. If the condition returns, they repeat the scarification. Nor do they neglect to burn the protuberances with their moxa, that is, the down of the leaves of the artemisia. Burning is a practice so common and accepted among the Japanese for all, even the most trivial, afflictions that they almost appear to have sworn obedience to the words of our own Hippocrates (whom they have never read). Hippocrates says, "If pain has located itself in any one place and settled there, and medicines have not been able to drive it out, cauterize the location of the pain, wherever it may be; cauterize with raw flax"(*On*

Affections 29).[9] But we will treat the topic of moxibustion later in its proper place.[10]

Notes

1. Pericàl, or maduromycosis or Madura foot, caused by any one of a number of fungi, is a chronic infection afflicting principally the foot and only rarely other parts of the body. The foot is enlarged and develops multiple sinuses and abscesses. Amputation or death from secondary infection are possible outcomes. The disease is endemic in widely scattered districts in India. Kaempfer's is the earliest description of pericàl.

2. The apostles are said to have divided the countries of the world among them, and India fell by lot to Judas Thomas (also known as the Twin). He declined to go because he was weak even after the Lord appeared to him. The Lord then sold Thomas to an emissary, named Hahban, from India. Thomas converted a number of Indians in the Madras area and is said to have been buried on St. Thomas Mountain just outside the city of Madras. See Bernhard Pick, *The Apocryphal Acts of Paul, Peter, John, Andrew and Thomas* (Chicago, 1909).

3. Coromandel: southeast region of the Indian peninsula.

4. Galle: a seaport on the southwest coast of Ceylon. Dutch interest in Ceylon was stimulated by a desire to control trade in the island's cinnamon, considered the best and finest. They captured Galle from the Portuguese in 1640 and built a powerful fort that still stands.

5. Omura: a fiefdom in southwestern Kyushu, Japan.

6. For Cochin, located on the Malabar coast, see Observation VII, note 7.

7. Marcus Aurelius Severino (1580–1656), *De abscessuum recondita natura libri VIII* (Naples, 1632). This work, which includes a section on breast tumors, has been called the first modern textbook of surgical pathology. Born in Tarsia in Calabria, Severino practiced law before turning to medicine and anatomy.

8. Probably the mango.

9. Hippocrates (469-399 B.C.), the most celebrated Greek physician. Unfortunately, not a single work ascribed to Hippocrates enjoys undisputed authenticity. Text and French translation are available in: É. Littré's *Oeuvres complètes d'Hippocrate, 10 vols.* (1839–61; facs., Amsterdam, 1961–62). *On Affections* is found in 6:206–71.

10. Moxibustion is the subject of Observation XII.

Observation IX
Snake Dances of Eastern India

I
Description of the Cobras de Cabelo

A shrewd and deceitful class of men whom we call tricksters, mountebanks, and quacks pervades, like a real nation of mortals, not only all regions of Europe but even those of civilized Asia. Some please their audiences with clever and delightful acts of agility; others entertain with trickery and deceptive illusions. To the latter category belongs the famous dance of a certain snake especially venerated in eastern India. The snake is called the Naja, but it had come to the attention of students of nature under its Portuguese name, Cobras de Cabelo.[1] Savage and exceedingly dangerous to humans, the beast injects its fatal virus with a bite. Persons bitten by the snake instantly experience heart pains and lipothymy;[2] unless an antidote is administered immediately, they die with convulsions.[3] If the antidote is administered even a moment too late, the person seldom escapes gangrene of the afflicted limb, which is almost impossible to cure.

Consequently, one discovers that the barefooted Indians are more terrified by and cautious of this type of snake than of any other. This viper (such is their classification) attains a length of three or four feet and a moderate thickness; it has a scaly and ornately striated skin, which is taut, rough, and dark but is white on the abdomen. When aroused, the snake has a peculiar habit—it inflates the skin on either side of the neck and expands it into a flattish or compressed orb like a wing or fan.[4] This fan has on its back the precise shape of an eyepiece delineated in white. The small circles on each expanded fan as well as the arc that connects these parallel circles are imprinted on the head itself.[5] When the snake darts forward to give battle, it rises up with body erect and proudly extends its fans, the indices of its anger, about its head. As gaping jaws display an array of deadly teeth, it strikes, unless cautiously repulsed, with wondrous speed at its foe. To manage

Plate 4. *Snake Dance*

this fearsome beast, not governable with halter or bridle, and to rouse it with song so that the spectator observes motions that rival those of a dancer—all this, those who only hear of it, judge to be magic or incredible, while those who actually see it, judge it an admirable performance. But if you study closely the nature of the snake dance and the method of training, all your admiration and acceptance of an act of magic vanish. I use the term dance or rather I misuse it, so as not to confuse the reader by the term leap. First, however, I will present an account of this dance.

II
Performance of the Viper's Dance

The trickster, usually accompanied by a colleague or a boy, is equipped with one or two vipers and traverses a region from village to village. He places his snakes on display at a crossroad or in the marketplace so as to gather a crowd (as usually happens in public places) for a public performance of the snake dance. Or he may even go from door to door, inviting the residents to a dance that, he proclaims, has never before been so charmingly and so curiously performed by a viper. When so requested, he fulfills his promises before the door or within the walls in this fashion. First, he produces a portion of a root, a supply of which he carries tied in a loincloth, while saying that by its protection he can approach the snake without harm and render himself immune to deadly strikes. After he has enclosed the little root in his right hand, he shakes the viper onto the ground from a container in which it is housed and mildly irritates it with a stick or the fist that he protected with the root. The viper suddenly turns toward its enemy, supports itself with the lowest portion of the tail, raises its body high, expands its fans into an orb, and vibrates its tongue repeatedly with a hissing sound as with gaping jaws and reddening eyes it timidly and tremulously stretches toward its master's fist. Then as the trickster begins to intone a ditty, he moves his bare fist in various ways and with a kind of rhythm from right to left and in turn from left to right, and sometimes up and down. The snake, its eyes fixed on its master's fist, imitates charmingly with its head and entire body the movements of the fist. With tail motionless, the snake's head turns this way and that way a distance of two spans from right to left and in so doing presents something resembling a dance lasting about one-eighth of an hour. For that period of time the viper can be worked; subsequently, when tired and annoyed it abandons the erect position and movement. To preclude weariness and flight, the trickster breaks off his song at the right moment, and the viper at once drops its body to the ground and ceases to dance. His task thus concluded, the performer collects a contribution or payment from the head of the household or the bystanders and returns the viper to its jar.

Now, you will ask what causes the viper to imitate the movements of the fist. Is it the hidden power of the little root or rather the enchantment of snake's master? The tricksters say that the two jointly exert compulsion. The root prevents the snake from doing harm, and the song makes it leap up. Accordingly, they attempt to sell the spectators the little roots and do not readily permit anyone who has not purchased a piece for protection to approach the dancing viper at close quarters. Indeed, they do not show the roots in their entirety (lest they be identified) but only in pieces one inch long. These resemble sarsaparilla in taste and external appearance but are a little thicker. But let us not be so demented as to believe that the root and song really exert an influence. I tested the power of the little root. I purchased two pieces from a trickster for a few coins, and despite the seller's protestations I threw them to a viper who was resting on the ground after completing a dance. The viper was not in the least disturbed by them; it made no effort to withdraw nor did it show any sign of aversion. I made the same test with garlic at the request of the trickster to whom I myself, not having been persuaded otherwise, had promised more genuine results in accordance with the principle of antipathy. The object, however, did not terrify the viper. I believe that the ash tree and the various plants, which Pliny and writers on natural magic claim put serpents to flight,[6] were not any more effective. In this century, no one but a fool is persuaded that serpents are charmed by song so as to dance.[7] Neither are insects captivated by song, nor does the passage of David in the Holy Scriptures appear to teach this.[8]

To be brief: it is fear alone and habit born of fear that induce that noble and docile beast to observe the chastising fist and intently follow its motion.

III
Practice and Training of Vipers

In the land of the Ganges I happened to observe the method of training at the residence of a Brahman outside the city of Nagapatana[9] in a rather remote suburban location. The Brahmanic rites that have increased in number over a long period of time and that have here and there been deprived of stipends by Mohammedan kings have driven this group to another way of life, as the returns from their temples no longer suffice for support. In Malabar Brahmans have fallen to the lot of carriers and enjoy a monopoly granted by their kings. Only those born of their holy stock may carry within and export merchandise from the provinces.

But to return to my man. He had devoted his life to taming vipers and to adapting them to various movements. Once subdued, the vipers were sold to tricksters. Now, this is a type of serpent that is at the same time both the most poisonous of all and the most intelligent and docile, if one may so use these terms. The Brahman kept individual serpents in covered earthenware jars whose size allowed the vipers sufficient space to coil the body. Training was conducted in the late afternoon (and perhaps in the early morning) when the sun, to be sure, is normally weaker. As the time approached, he arranged outdoors on the sandy ground as many jars (I counted twenty-two) as he possessed. Each serpent in turn was removed and trained for a shorter or a longer period depending upon its progress, and at the end was again enclosed in its jar. Some were new or barely started on executing the movements, while others were more manageable and experienced.

This was the method of training. Once the cover was removed, the jar was inverted and the viper shaken out. When the viper began to flee, the trainer turned it toward himself with a stick. Thus challenged to fight, at which time the serpent usually makes its stand willingly, it sprang to bite, but the trainer had snatched up the jar and thrust it out like a shield.[10] Struck on the nostrils and frustrated in the attempt to injure, the serpent sprang back. During this struggle the trainer kept the viper erect for a long time (some for one-half and others for one-quarter of an hour) until, again and again baffled by its injury, it refrained from springing. Yet all the while, with fans extended and fangs bared it did not cease to fight and cautiously followed the motion of the offending shield at which it stared with glowing eyes. The trainer was careful not to allow the exercise to injure or tire the viper excessively, for otherwise he would have crushed its spirit and rendered it timid and inactive with the result that, when summoned to the arena in the future, it would undertake flight rather than combat. So managed day after day, the vipers gradually accustom themselves to rise up before the thrusting of the jar and, when at length the jar is dispensed with, before the thrusting of the fist alone. With heads proudly extended in fans, they imitate the movements of the thrust fist this way and that way, quickly and slowly and with a kind of rhythm, ever fearful, as it seemed to me, of being injured by the object. If the trainer accompanies with song these graceful movements first born of combat, a not unpleasant spectacle arises, which among the ignorant herd has acquired the famous title of snake dance.

IV
The Method of Removing Venom from Vipers

We have intimated that it is vain to desire the protection of little roots against vipers. Why, then, do tricksters hour after hour expose themselves to danger for a few coins? The parasite Damocles would refuse even for the king's delight.[11] No wonder they remove all the poison from vipers prior to a performance so as to manage them without injury, for it often happens that a serpent bites an incautious master, but without venom the small wound is trivial. All the virus that the serpent has is located in the upper gum. It is nothing but the salivous fluid that is deposited from the head into that part of the gum situated at the aforementioned canines within the oral cavity. As the gum is compressed in the course of biting, it cannot be otherwise but that the virus is introduced at the same time the small wounds are made. Accordingly, to elicit this saliva from its container, they throw a cloth or anything porous and soft to the vipers who have previously been annoyed. This object, when bitten by the viper several times, imbibes all the saliva from the gums. Some first grasp the neck with their fingers and press it so that the viper, anxious to strike back, releases the virus. When the serpent has been provoked in this way, only then do they throw something to it to cleanse away the remaining liquid. The tricksters cause such an emission of poison each day or on alternate days, being careful, however—and this they keep as a dark secret—not to allow the despoiled serpents to feed on fresh grass or anything green, for otherwise with this nourishment the serpents would replenish themselves within a few hours.[12]

I say nothing about the peculiar power of this venom, since the indefatigable investigator of nature's truths, Redi (Vol. 2, *Opusculorum Observ. de Viperis*), has already treated the matter so very learnedly.[13] I only wish to add a unique account in support of the position of that most distinguished Italian and against the position of the equally distinguished Frenchman, Charas.[14] The former has located all the serpent's poison in the saliva; the latter, in the idea of vengeance.

In the year 1689 I spent a few months in Batavia, the Dutch capital in the Indies.[15] One Sunday, an enlisted soldier, a German, who had studied pharmacy in his homeland, was standing guard at the fortress gate named after the water. I greeted the man before the assembly during which the young man met an unexpected death from a viper, although it was itself dead. A comrade had killed the serpent one half-hour earlier not far from the gate. The young man had arranged to obtain the snake's body while on guard duty, as he was anxious to cover the scabbard of his sword with the skin of the snake. Using his teeth, he took hold of the snake's head and with a knife cut the skin around the neck, as one does with eels. Scarcely had he begun to strip the loosened skin, when suddenly he fell dying. As the gates were closed for the assembly and there was no one to assist with a remedy, within a half-hour he died of convulsions. I have no doubt that the skin on the tip of the man's tongue or on his lips had recently been broken and was then brought into contact with the viper's mouth still dripping with saliva.[16] The venom, absorbed by the capillary veins and carried to the seat of life, occasioned dire symptoms and death itself.

Everyone in this country knows that viper's saliva, when drawn in by a slight and still bleeding wound, however tiny it may be, even if only a needle prick, can cause death, while if a person swallows even a large amount of venom, he will suffer no harm at all. They do not hesitate to assert the same about the venoms of other animals (scorpions, spiders, wasps, etc.). Exception is made in the case of tiger's whiskers, which the Indians, once the beast has been killed, are obliged to bring to their rulers as the strongest venom for internal use and one thus far not commonly known.[17] But what intention to harm or what idea of vengeance could there have been in our viper who was quite dead? I am equipped with more experiments, which I would add if this were the place to defend Redi's position.

I have also seen performers scurrying about to earn some coins with serpents of great size, thick, longer than six feet, and hanging down from shoulders to the ground. These serpents also are part of their show. With hands wrapped with bandages they violently break the serpent's neck and then insert a linen cloth two or three times between the teeth of the upper jaw to wipe off all the salivous fluid. Next, they stick out their tongue as far as possible and insert it into the mouth of the serpent. As they withdraw their tongue, they inflict harmless wounds with the points of the serpent's teeth. The wounds, long, bloody stripes, are displayed for the spectators' contemplation. This sluggish and crude type of serpent—it was smooth, whitish and variegated with large blackish patches—is said to have little venom and great strength. To the same class belong the serpents that in the native language are called constrictors. They are so named because they are said by some sort of agility to seize, encircle, and constrict quadrupeds to death, and then finally to swallow them down after suffocation—even oxen, as the distinguished Cleyer (*Ephem. Germ.*) reported, though I have not found evidence for this or heard of it from trustworthy people.[18] It is nonsense and a myth of the common people, just as are the stories about the Roc, which is said to be a bird of great size and strength on the island of Madagascar and capable of carrying off an elephant.[19] Not even the name, however, is known on that island. Those who believe such things are fools; those who demand belief are scoundrels!

Notes

1. The term *cobra de capello* (spelling varies) is Portuguese for "hooded snake" and reflects the early presence of that nation in India. Considerable material on the cobra had been published during the previous century by the physician and naturalist Garcia da Orta (see Observation V, note 22).

2. lipothymy: fainting or loss of consciousness.

3. The cobra's venom acts rapidly and directly on the nervous system. Before the advent of modern antivenines, the only effective treatment for a bite was the immediate excision of the site. So-called snake stones and other "cures" were of no avail.

4. The lateral dilation of the neck is effected by the cobra's raising and pushing forward the long anterior ribs as the elastic skin is stretched taut over this framework.

5. The spectacle markings are quite distinct on some cobras while not visible on others.

6. Pliny the Elder writes:

Indeed the leaves of the ash are found to be serviceable as an exceptionally effective antidote for snake bites, if the juice is squeezed out to make a potion and the leaves are applied to the wound as a poultice; and they are so potent that a snake will not come in contact with the shadow of the tree even in the morning or at sunset when it is at its longest, so wide a berth does it give to the tree itself. We can state from actual experiment that if a ring of ash leaves is put round a fire and a snake, the snake will rather escape into the fire than into the ash leaves. By a marvelous provision of Nature's kindness the ash flowers before the snakes come out and does not shed its leaves before they have gone into hibernation. (*Historia Naturalis* 16. 24. 64)

7. Snakes are unable to hear airborne sounds. They follow not the sound but the movement of the musical instrument and the charmer.

8. 1 Sam. 16.14–23, where reference is made to the power of music to exorcise King Saul's evil spirit.

9. Nagapatana (Negapatam) is a seaport on the Coromandel coast of India. Settled by the Portuguese, it came under Dutch control in 1660.

10. The cobra's movement when striking is quite slow when compared with that of certain other snakes: the cobra strikes with only one-sixth the speed of

the rattlesnake. Hence, an experienced trainer is less prone to being bitten while intercepting the cobra's bite with a jar.

11. Cicero relates the famous tale of the sword of Damocles. The courtier Damocles, with the excessive zeal of a professional flatterer, claimed that no mortal was happier than King Dionysus. The king then allowed Damocles to sample this happiness with a couch of gold and a table lavishly set, except that a sword was hung over the head of Damocles by a horse's hair. Damocles begged to be allowed to depart. Cicero concludes by asking whether Dionysus made his point that nothing gives pleasure to a man threatened by death (see *Tusculan Disputations* 5. 61–62)

12. This technique is termed *milking*. For an interesting discussion of the methods employed by snake charmers to render snakes harmless see James A. Oliver, *Snakes in Fact and Fiction* (New York, 1958), 95–118. Defanging appears to be a common practice and to explain the celebrated Hopi Indian snake dance. American carnival performers speak of their snakes as "fixed" or "hot"; venomous snakes are regularly fixed.

13. Chauncey D. Leake assesses the contribution of Redi's *Osservazioni intorno alle vipere* (Florence, 1664):

> It was not until the 17th century that systematic studies of venoms were made in a manner that may be regarded as scientific. In 1664, Francesco Redi (1621–1697) wrote the first methodical work on snake poisons. Redi demonstrated that in order for the snake poisons known at that time to produce their characteristic effects, they must be injected under the skin. When taken by mouth, certain venoms were clearly recognized as harmless. ("Development of Knowledge about Venoms," in *Venomous Animals and Their Venoms*, eds. W. Bücherl, E. E. Buckley, and V. Deulofeu [New York and London, 1968], 8)

First Physician to Grand Dukes Ferdinand II and Cosmo III of Tuscany, Redi is also credited with the first significant attack on the theory of spontaneous generation (see also Observation IV, note 13). Kaempfer refers the reader to vol. 2 of *Opuscoli varj di F. Redi* (Florence, 1684–91).

14. Moses Charas (1618–98), *New Experiments on Vipers* (London, 1670). The French edition, *Nouvelles expériences sur les vipères*, was published at Paris in 1669.

15. Batavia: city and seaport in Java.

16. This is quite possible. Reflex action by the dead snake could also have inflicted the fatal bite.

17. William Baze observes, "But most of the tiger-skin rugs one sees have been given false whiskers, the originals having been stolen by natives before

the animal was skinned. There is a popular belief that a tiger's whiskers, treated in a certain way, make a violent poison to which there is no known antidote. A hunter who is anxious to keep trophy intact is therefore obliged to watch it carefully if he wants to prevent this highly regrettable depilation!" *(Tiger! Tiger!*, trans. H. M. Burton [New York, 1966], 18).

18. Andreas Cleyer, physician of the Dutch East India Company, published a large number of observations in the *Ephemerides* of the German Academia Caesarea Naturae Curiosorum. Among reports by travelers is the following: "There are some serpents both in Asia and America of monstrous bigness, 25 foot long; as was that, the skin whereof is kept in Batavia, which had swallow'd a Maid of 18 years of age" (*The Travels of Monsieur Tavernier Bernier*, trans. J. Phillips and E. Everard [London, 1784] 1:155).

19. The roc (rukh) is a fabulous bird said to be large enough to carry off elephants. Legends of the roc are found in E. W. Lane's *Arabian Nights* (1839), and H. Yule, *The Book of Ser Marco Polo* (1871), 3.33. Huge fronds of the raffia palm from Madagascar were thought to be the bird's feathers. In the *Arabian Nights* (chap. 20, "The Fifth Voyage of Es-Sindibad of the Sea"), the merchants arrive at a large, uninhabited island and break "an enormous white dome," which turns out to be a roc egg. Mother and father rocs pursue the ship and demolish it with a mass of rock from a mountain.

Observation X
Two Indian Antidotes

I
The Mungo Plant

The boiling blaze of the tropic sun intensifies and sharpens the salutary powers of nature's resources. But the same effects also apply to the deleterious energy of poisons. In cooler regions solar influence is weakened. I need not cite more than two examples. Who would not be amazed at the deadly swiftness of the sap from a tree of Makassar? Weapons dipped in this substance need only the slightest contact with the blood to extinguish life instantly as if it were a lamp. Who would not shudder at the venom of the Naja snake that the Portuguese call cobras de cabelo?[1] Its slightest bite kills. Just as the Indians dread these two poisons more than any others, so two antidotes, which specifically counteract these poisons, enjoy the highest repute. My intention is to discuss briefly these antidotes.

The first antidote, which comes from the vegetable kingdom, is the root of a plant that the Malays call Hampaddu Tanah, that is, earth bile. Its name is derived from the extremely bitter taste of almost every part, especially of the root, which resembles the intense bitterness of bile. The local Portuguese call it Raje, or Mungo root, after a variety of weasel or ferret;[2] the Indians, Mungutia; the Portuguese in this area, Mungo; and the Dutch, Muncus. Garcia da Orta calls the plant Quil and Quirpele, after the animal whose habits are believed to have first drawn man's attention to the root and its use as an antidote.[3] A natural hostility inclines the mongoose to attack a snake, even as a cat is inclined to pursue a mouse. It is commonly believed that if the mongoose is overcome by the cunning or strength of a snake and sustains a bite, he will break off the fight to search out this root as an antidote and that instantly restored by eating it he will renew the struggle. Let the natives believe the story! I have discovered this about the mongoose. When bitten by a snake or tired by the fight, the

96

mongoose disengages himself from his victorious opponent and scurries from the arena of combat into the fields to feed on the small roots of herbs. Having quickly regained his strength through the food, I would judge, he again prepares to renew the contest with his enemy, if the enemy is still present. The mongoose (which the illustrious Ray incorrectly differentiates from Garcia's Quirpele, *Synop. Anim. Quadrup.*)[4] has the shape of a squirrel but is slightly larger and slower. It is adorned with grayish fur, which is longer around the tail and spotted with black for decoration. When fed, it is readily domesticated. I had one that slept with me and followed me through the city and country like a small pet dog. The hot countryside of all Asia as far as the Ganges, even in those regions where the aforementioned root never grows, supports the mongoose.

Let us return to the virtue of the root. The Indians, especially the inhabitants of Java and Sumatra, whether they learned its use from the mongoose or discovered it by chance, regard the root as a proven antidote for poison that has entered through a wound or was taken internally. The small root enjoys the highest repute from the superlative power it brings to bear against the Makassar poison, that deadly one scarcely susceptible to any other remedy. This poison is a milky, rich sap that is collected from a certain freshly wounded tree. The tree grows in the remote forests of Celebes, especially in the province of Turasia.[5] The natives call it Ipu; the Malays and Javanese call it the Upa. There are said to be three varieties or species of the tree. They differ only in the degree of deadliness of the sap and its color (dark, reddish, and whitish) when it dries.

Those undertaking the collection of the true sap on this tree, which only grows in Makassar, must run the gravest risks to their lives, for when searching for the tree, they must penetrate places thick with thorns and beasts. When the tree has been located, it will instantly suffocate its attackers with an outburst of fumes, unless the tree is wounded from a distance and from the direction from which the wind is blowing or the breeze prevailing. Birds flying about a recently wounded tree are also said to experience its great power. The collection of this deadly liquid is entrusted to those condemned to death for the crime of witchcraft. If the criminals return with the liquid, their lives are spared. Having escaped Scylla only to encounter Charybdis, they undertake this task with extreme caution and circumspection. No wonder they enter the forest equipped with a long reed of the large and stout type with which they build their homes. The reed is the hardest type, as thick as the shin or ankle, hollow, and geniculate with internodes at intervals of a foot or longer. They finely sharpen it at either end so that the reed can penetrate the bark of the tree. When the tree has been sighted, they attack it from afar and from the direction I cautioned. Positioned as far as possible from the tree, they thrust the point of the reed hard into the tree and gather only so much of the liquid flowing from the wound as can be contained in the hollow of the reed up to the nearest joint. Laden with these spoils, they withdraw into the oncoming wind, and soon safely located beyond the exhalation of the wound, they pour the liquid into a glass jar with equal caution and cork it. With all punishment and danger now past, they return to present the king with the baleful price of their lives. Such is the story told me by Celebes citizens, now called Makassars. But what truth unembroidered with fiction can one learn from the mouths of Asians?

It is a fact that the king of the Makassars and the other chiefs of this realm dip their spears and knives (their only weapons) in this deadly sap. Unless the weapons are properly covered, in time the poison weakens from the circulation of air. This is less likely to occur if the poison adheres to tiny cracks in the weapons where it is tightly packed. The famous weapon is a small dart cut from a reed. Its very fine tip encloses a channel smeared with poison. The soldiers of this country are usually armed with such weapons. They send the darts flying against the enemy by blowing through slender, arundinaceous tubes (the sort our boys employ to attack small sparrows). The lethal dart plants itself in the body—a terror to the unclothed people of the Orient but harmless to soldiers dressed in European fashion.

I have heard that other people attempt to make poisonous weapons even with the venom of serpents. But the people of the Orient find its collection quite laborious and the poison less powerful because when stored, the salivous liquid dries and loses most of its deadliness. The collection of venom for poisonous weapons is said to be routine among the Africans who live near the Cape of Good Hope. In that region is found a certain type of snake of moderate length and variegated with a golden color. Behind the canines it has prominent sacs swollen with venom. After killing the snake, the natives cut out the sacs and preserve the fluid for use as a poisonous dip. I have not actually seen the snake, but I have seen a skillful picture of it in lifelike colors along with pictures of other local creatures at the home of the noble and revered van der Stel, the Dutch governor on the Cape of Good Hope, a gentleman of great curiosity and exceedingly erudite in the culture of this region.[6] I have no doubt that weapons can be dipped in other drugs. The effects of tobacco oil are well known. I myself have not tested the power of garlic juice or the Spanish poison prepared by cooking hellebore roots.

But let us return to our little root. One takes it both as a preventive and a curative for poison whether anticipated or already administered. The dose is one or one-half drachma with a drink of sweet water or any liquid. The root is reduced to a powder by rubbing it on a rough stone or by a more careful method. They advise, not unwisely, placing on the afflicted spot itself a poultice made from the powder of the root and spittle or water. Before the Europeans—as either friends or enemies of the people of the Orient—learned of this bitter root, it is said that the natives revealed no other antidote for the fierce poison but dried human excrement. The wounded person had immediately to swallow down a quantity sufficient to cause violent vomiting, which would overpower and thwart the inflicted virus. The process is one of tearing out, eliminating, and the like. Armed with this excremental antidote, they joined battle with the dark people of Celebes. They knew only this disgusting means of saving their lives—a mockery—until it became clear that the natives cherished another antidote, whose identity could be extorted only by torture. Some people solemnly assert that if the sap is fresh and delivered by a recently smeared weapon, the victim's life cannot be saved by any antidote or by excision of the site of the wound. The victim falls dead before anyone can extend a helping hand. Upon leaving Java, I brought back a supply of the root, which I am employing with good results on wasting fevers and malignant diseases. On the basis of many successes,

fitanis impofito, adverfus viperarum morfus præftat au-
xilium, externè applicatus.   In ferpente, quòd vulgò cre-
dunt, non invenitur, fed arte fecretâ fabricatur à Brah-
menis.   Pro dextro & felici ufu, oportet adeffe geminos,
ut cum primus veneno faturatus vulnufculo decidit, alter
furrogari illico in locum poffit; Proinde poffidere indi-
genæ nifi geminos cupiunt; quos cum goffypio capfulæ
inclufos probè affervant.   Aliam obtineo ophitidem, in-
ftar filicis duriffimam ponderofamque, uncialis longitu-

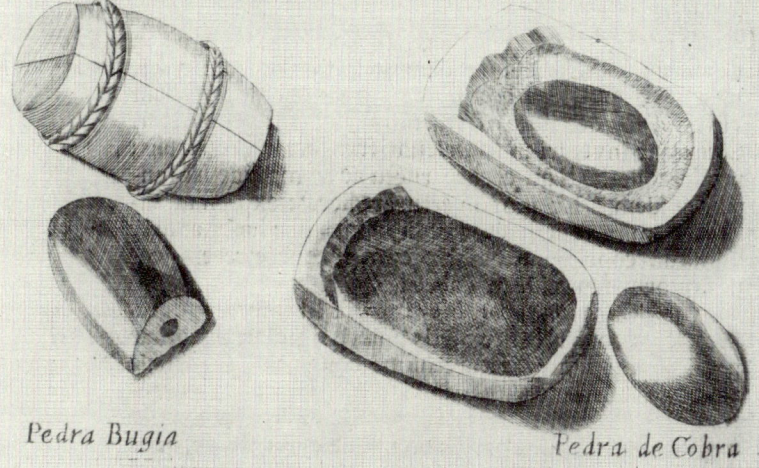

Pedra Bugia

Pedra de Cobra

dinis, figuræ quodammodo tornatilis, & quafi ex annulis
compactæ; quales inveniri in capitibus maximorum fer-
pentum credulus perfuadetur vulgus: ego apophyfin
alius petræ, vel in ipfâ petrâ molliori genitam judico, velut
lapides bufonius, aëtites, lyncius, oculus catti, gloffo-pe-
tra, & fi quos alios vulgi error tranfcribit ex animalibus.
                                                            Fidem

Plate 5. *Snake Stone*

I assert its infallible power against the bite of a rabid dog in both animals and humans.

Our plant, named by the Indians from its bitterness and by the Portuguese after the ferret, grows as far as I know, on the islands of Java, Ceylon, and Sumatra, and perhaps, as I suspect, in other regions. It grows to a height of a foot or more, and viewed from a distance, it resembles the *centaurium minus*.[7] Its root is single and simple, a span in length, as thick as a finger, and twisted in many a winding turn; the firmly adnate bark is fungous, red, wrinkled; the substance is lignose, hard, without fibers, fragile, whitish in color, and very bitter in taste, like *Gentiana*[8] but more delicate and less disagreeable. A straight, slender, and even stalk arises with widely separated branchlets on alternate sides. The branchlets are adorned with leaves arranged in the same pattern, but more widely separated, narrow, an inch long and lance-shaped. Single styles as long as the width of two or three fingers, slender and deciduous, arise from the furcate axils. On the styles, there are florets, about fifty, more or less, very small, distributed in several fascicles, and conglobate in an umbel small enough to be covered by a finger nail, and beautifully red. Under a microscope they are shown to be bare and tubular with an extremity divided into five or six lacinia. As the floret dries, its seed protrudes from a very tiny umbilical pedicle. In some cases there is a twin growth, that is, from two seeds united at the bases with divaricate tops. In other cases the growth is one and simple, unequally round, somewhat conical from a kind of compression, bony, the size of coriander seed, covered with a skin that is fleshy, adnate, wrinkled from dryness, and of a color indiscriminately blackish or pale green. The interior nucleus is white, oily, and rich, with a vapid, sub-sweet taste. The Mungo plant, or *Phaseolus orthocaulis* described by Fabio Colonna[9] and published by the distinguished Ray (*Hist. Plant.* LXVIII, section 3, chapter 6)[10] is unknown to me.

II
The Snake Stone

The second antidote is a stone that comes, so they believe, from the animal kingdom. When applied externally to small, poisonous wounds, it clings with very great tenacity until—the words are those of the natives—saturated with venom drawn from the wound, it falls off of its own accord. While it does furnish superb assistance for every venomous wound, whether the poison be inflicted by sting, insect bite, or by a deadly weapon, it is, however, considered the specific and infallible antidote in humans against the bite of the venomous snake, the cobras de cabelo. To effect properly a cure, two stones should be kept in readiness so that when the first falls off, the second can immediately be applied and drink in the remaining venom. If, however, only one stone is available and it indicates by spontaneously falling off that venom is still present in the wound, the stone should immediately be immersed in fresh milk. Milk instantly washes out the virus from the pores so that the virus, variegated with yellowish and bluish colors, can be seen floating on the surface. When thus freed from the venom and diligently cleansed, the stone should once again be applied to the wound. They say that if the stone no longer falls off but must be pulled away by hand, this is a sign that the venom has been completely drawn off, for the porous stone, not yet filled with venom, can no longer be saturated.

Much discussion, research, and disputation are needed with respect to the efficacy of this stone and the experiments daily conducted with it in India. Be that as it may, as many Hindustani as I have questioned about the stone avow the truth of its efficacy. And I have found countless Hindustani who swore to me by the Stone God that they themselves had conducted tests and that with their own eyes they had seen tests made. Consequently, among the Indians it is a crime of unpardonable skepticism to entertain doubts about the virtue of the snake stone against the bite of the cobra. Of the many I could cite, I prefer to include one exceedingly rare case that was made known to me by the noble and distinguished de Pavillon when in the course of conversation I expressed my doubts about experiments with snake stone. The incident occurred at the home of de Pavillon while he was governor of Coromandel[11] and his whole family was eyewitness to it. One of the domestic servants had been bitten on the foot by a cobra and was screaming with the most severe pains. Immediately, the stone was produced and placed on the bite. When it fell off, there was available no other stone nor any sweet milk to cleanse the pores. A wet nurse, fearing the swift action of the ailment, gave milk from her own breasts upon the stone. As a result, however, her nipple at once began to ache, and the breast began to swell and become inflamed. She tossed on the brink of death for three days, and by the tenth day she barely recovered from a hardening of the breast. Since the skin of her nipple had been a bit damaged by the nursling's gums, small veins were opened that admitted the poisonous pollution drawn out by the warm milk.

But how shall we reconcile the experiments of the Indians with the many conducted by the famous Redi, which yielded conflicting results (they are cited in *Opusc.*, vol. 2)?[12] Supporters of the stone will, I think, deny that the experiments of Redi are valid to disprove the efficacy of this antidote, since his experiments were conducted with animals and not with humans, with tobacco oil or various snake venoms and not with cobra venom, and finally in a climate colder than that of torrid India. As a reporter, I make no judgment, but I do frankly confess that the value of the stone always remained questionable in my eyes as long as I had not by my own experiments investigated the possibility of hidden error and deceit. Granting that I have nothing to contradict the many who extol the stone and boast of experiments, nevertheless I have never dared to test the value of the stone in critical cases that arise, as they usually do, quite unexpectedly and with instant peril to life, because I could not grasp how the snake virus, if it is as fast-acting as everyone says it is, can be withdrawn from so large a quantity of fluid by the adhesion of a stone. But if the virus is not so fast-acting (as I at least seem to have discovered in many illnesses) but requires a kind of delay before, diffused and dissolved in the wound, it can be completely absorbed by the capillary veins and carried to the brain and vital organs, then it will be permissible to gather up earthenware, spodium, ashes of stag antlers, and alkalis of a similar type that likewise adhere and can perform the same functions as the stone. When applied to the wound, they can catch and drink in equally well through their pores the contaminated fluids and along with them the venom. Consequently, I have more prudently decided to take refuge in my venerable haven, theriac.[13] First, a tourniquet is applied on the limb above the wound, then the wound is scarified; next, blood is forced out; finally, I rub on and apply theriac while a dose of the same electuary is taken with water as a sudorific. With this treatment and without the stone, a cure has always followed rapidly enough in cases of fresh wounds.

The snake stone has a somewhat ovate and compressed shape, is longer than one inch, wider than one half-inch, sometimes rather gibbous on one side or the other, as thick in the middle as a goose feather, with a thinner edge, a smooth and bright surface, and an unpleasant color, which is blackish and partially whitish. Its substance is firm and hard, but light where it is porous and is somewhat corneous, so that the stone appears to be formed from stag antlers softened and dyed in steam or some fluid, unless perhaps it is a fragment of variegated Lapis Cananoor. (This stone is so named from its source, the district of Cananoor in Malabar. The Portuguese there call it Pedra frigue from its cooling effect. There are three kinds or colors, to be sure: white, citrine, and dark blue, the last of which is very like nephritic stone in all respects but lightness.) As many as I chanced to see on the Indian mainland and islands were of the aforementioned condition and shape, clearly different from those acquired by the Reverend Fathers and illustrated by the distinguished Redi. Whatever their shape be, on first sight it will be obvious that they are quite unnatural and were not formed in the head of a snake, as is commonly believed. Consequently, those who search for them in heads of serpents do so in vain. Two stones, enclosed in wool, are usually kept in a single capsule fashioned from the terete and light cork oak, so that their owner may conveniently carry them without danger of breaking. By no means are they everywhere offered for sale, but when they are, they are considered rare and command a high price. The identity of their makers is not known. Many, however, suggest the religious among these heathen, whom they call Sjoges, or priests, and Brahmans. See the illustration in fascicle II, report IX, section 3 [Plate 5].[14]

Notes

1. cobras de cabelo: hooded snake (see Observation IX).

2. The mungoose (mungo) plant, *Ophiorrhiza mungos*, of the family Rubiaceae. Snakeroot is the general name for numerous roots or herbs considered antidotes in regions where snakes abound.

3. Garcia da Orta's *Coloquios dos simples, e drogas he cousas medicinais da India* (Goa, 1563), 1.44. Kaempfer refers to the mongoose.

4. John Ray (1627–1705), called the father of English natural history. Kaempfer's reference is to *Synopsis Methodica Animalium Quadrupedum et Serpentini Generis* (London, 1693), 197.

5. Turasia (Toradja) is located in the central Celebes.

6. Symon van der Stel (1639–1712), first governor of the Cape of Good Hope, was noted for the extraordinary interest he took in the country: founding Stellenbosch (1680); opening the Drakenstein district (1684); prospecting for copper; attending to tree planting, wine industry, schools, and so on. In 1692 van der Stel was also named Extraordinary Counsellor of India. His son Adriaen succeeded him as governor of the Cape of Good Hope in 1699.

7. *centaurium* (centaury): a plant whose medicinal properties were supposedly discovered by the centaur Chiron. Two species of the plant were distinguished: *centaurium major* and *centaurium minus* (greater and lesser centaury).

8. The bitter root of the genus *Gentiana* is employed in medicine to stimulate digestion and promote nutrition. The flowers may be blue, yellow, white or red.

9. Fabio Colonna authored *Phytobasanos, sive plantarum aliquot historia. . . Accessit etiam piscium aliquot, plantarumque novarum historia* (Naples, 1592), the first book produced with copper engravings for plant illustrations.

10. John Ray, *Historia Plantarum*, 3 vols. (London, 1686–1704).

11. Founded in 1664, the French East India Company established its first Indian factory at Surat in 1668. The site of the modern Pondicherry on the Coromandel coast, acquired in 1673, became the headquarters of the French East India Company in India.

12. Francesco Redi (1621–1697) argued that snake poisons were harmful only if injected, and some poisons could even be safely ingested; see Observation IX, note 13.

13. theriac: a compound of various ingredients used as a remedy against a

wide range of disorders, especially bites of poisonous animals. The most famous is *theriaca andromachi* (Venice treacle), a mixture of sixty-four drugs prepared with honey and reduced to an electuary. It was prescribed as an antidote for poisons.

14. Chapter 9 of fascicle 2 is titled, "Memorabilia of Mount Benna in the Persian Province of Laar." Section 3 treats various bezoar stones, their supposed medicinal qualities, and the animals from which they are derived. Kaempfer rejected current reports that *pedra de cobra* (cobra stone) was obtained from snakes. As for the potency of the stone, Kaempfer believed that the only virtues of *pedra de cobra* against snake bite were actual coldness and absorbency.

Observation XI
Acupuncture:
A Cure for Colic Employed by the Japanese

I
Description of the Disease

Colic, which the Japanese call Senki, is endemic in the islands of the densely populated empire of Japan. Indeed, colic is so prevalent that there is scarcely one adult in ten who has not at some time or other suffered from it. The air (otherwise most healthful), the local water, the food, the drink, and the native way of life all conspire to promote this disease. The unpleasant malady also attacks foreigners who eat the native food. I discovered this, to my grief, when upon arriving in this region we attempted (with cold local beer called Sakki)[1] to wash away the memory of the suffering that the angry gods of the sea had inflicted on us. The practice is traditional for travelers coming from the high seas. Sakki is brewed from rice to the strength and consistency of Spanish wine. Its nature is such that one should never drink it cold, but should sip it moderately warm from a dish, the way the natives do.

The term Senki, however, is not applied to any stomach pain, but only to the one that while tearing the intestines simultaneously causes convulsions of the groin. No wonder. For it is characteristic of either the nature or the intensity of this colic to bite the muscles and membranes of the abdomen. If one is looking for the cause or the substance of the pain, the Japanese hold the opinion that both with respect to the malady in question as well as all stomach torments, the cause is not lodged in the intestinal cavity (except perhaps in the case of a rather slight affliction) but adheres to the membranous substance itself of some part or other: the abdominal muscles, the peritoneum, the omentum, the mesentery, or the intestines. After stagnating there, the diseased matter turns into a vapor, or as the Japanese say, a very sharp spirit that distends, erodes, and tears the membranes. Conse-

108

quently, if the area in which the eroding spirit is imprisoned is penetrated and the spirit is set free from the confined area that it distends, the fierce sensation of pain caused by the enlargement will cease instantaneously. Westerners incorrectly term the disease colica, a Latin word derived from the word for intestine, which is usually not at fault. The gymnosophists,[2] however, in keeping with the better judgment of the Japanese and Chinese, prefer to call the ailment in their language a spasm of the abdomen and intestines.

This endemic disease of Japan has the following unusual symptoms. By convulsing the entire region from the groin to the false ribs or mucro, it very often causes the patient to suffocate as if he were suffering from hysteria. Indeed, after the disease has raged for a long time, it ends with tumors growing promiscuously about the body. Sometimes the disease produces the most hideous consequences: in men a large swelling of either testicle which usually suppurates in a fistula; in women, a large number of tubercula growing on the labia of the anus and vulva, normally accompanied by the loss of hair. This sarcocele (the Japanese have special names: Sobi for the ailment; Sobimotz for the patient) and these growths, however, are equally common and endemic in Japan from causes other than colic.

II
Description of the Needles

Before we ready our needle against this disease that is the enemy of that distant land, we should note the following. In the Orient, there are two sacred sources from which the healthy and the sick, physicians and quacks, rich and poor constantly seek surgical remedies in order both to restore and to preserve good health. The three nations of Korea, China, and Japan, which cling most tenaciously to tradition, solemnly assert that these sources were celebrated in their region from the earliest times long before the discovery and employment of medicine. What are these sources? Their very names are frightening! The first is fire;[3] the second is the sharp point of metal.[4] We are, however, not talking about the sharp point of deadly steel dripping with blood, nor about the fire of glowing iron. Occidental surgery, with its grim attitude toward human beings, prescribes the savage employment of such instruments against men. This approach is an abomination for those whose hearts are guided by human kindness and compassion. The Oriental fire, however, is agreeable and no more to be feared than that which the very gods of the region are pleased to accept as an offering. We are, then, talking about a gently burning cone rolled from the plant named after Queen Artemisia.[5] The metals employed by the Japanese are, of course, the most outstanding ones— the ornaments of royal palaces, sheer energy, the creations of the sun and moon through fire, and pregnant with the power of these two celestial bodies, as the philosophers would have us believe. Why prolong the suspense? I am referring to gold and silver! From these metals the local craftsmen fashion highly polished needles ideally suited for puncturing the human body. These needles are considered so precious that they are among the portable treasures that the Japanese, a nation with a love for boxes, carry about close to their hearts (their usual practice with all sorts of highly valued utensils).

The execution of these surgical remedies is so critical that a knowledge of the locations that are to be punctured or burned constitutes a unique art. The masters of this art are called Tensasj, or touchers, which means searchers for the proper locations, since the selection of the locations is fundamental to the entire operation. Those who work with the needle in hand are called Faritatte, or acupuncturists, whether they operate according to their own discretion or on instructions from a toucher. Since driving a needle into the human body involves a serious risk, a needle used for acupuncture must be very thin and made from gold or silver, which is very pure, entirely free from copper, and ductile. A unique art is required for tempering needles, since a needle that is not firm enough will ruin an operation. Not everyone knows the art, and of those who do know it, only persons who have received a license with an imperial seal are permitted to manufacture needles.

There are two types of needles. The first type, which may be made of gold or silver, somewhat resembles a writing instrument in shape, although not in size. I have in mind the quills our schoolboys use to spell or the styli the Indians use to write. Needles of the first type are four inches long and moderately thin, end in a very fine tip, and have

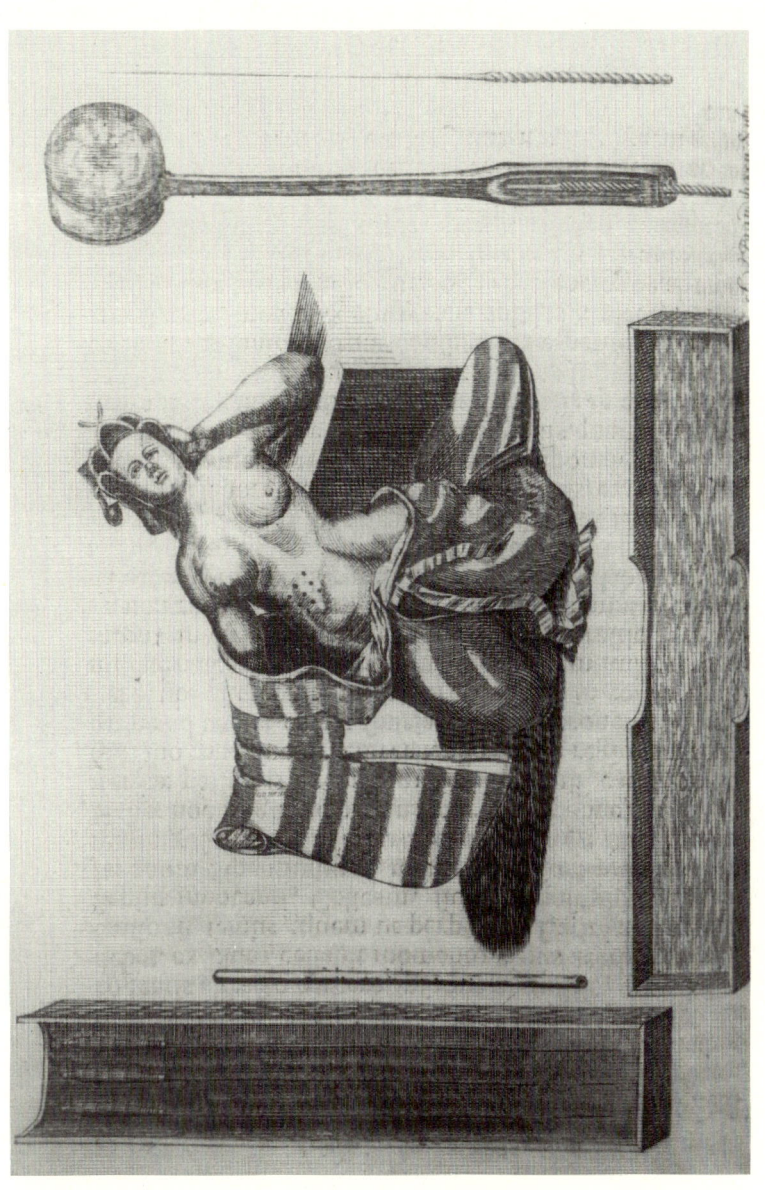

Plate 6. *Acupuncture Needles and Acupuncture Points to Cure Colic*

a spiral-surfaced handle with which they can be turned. They are not stored in a box but in a small hammer so designed that each side of the handle can hold a single needle [Plate 6, right]. The hammer is made of highly polished wild ox horn. The hammer is a little longer than the needle and has a rounded and compressed head. The side intended for striking has lead enclosed to furnish weight and is covered with violet-colored leather in order not to recoil when striking the needle. Needles belonging to the second type are always made of silver and are little different in design from needles of the first type. They have the same length as the first type, the thinness of a lute string, and a short, thick handle furrowed lengthwise. Several (the number is uncertain) needles are stored in a rectangular wooden box whose exterior is varnished and whose interior is covered with a loosely woven cloth [Plate 6, left]. The needles are inserted slightly into the tufts of wool. For the reader who is curious about names, both types of needles and in general all needles employed in this surgery are called Uuts barri, which means twisting needles. In addition, the second type of needle has a special name, Fineri barri, which means the same as Uuts barri. If a tiny brass tube is joined with it for an operation, as frequently happens, the tubed needle is called Fuda barri. The tube is slightly—about one-third of an inch—shorter than the needle and as thick as a goose quill. The tube is so designed to guide the needle to the precise point to be punctured and to preclude error during an operation.

III
Procedure for Acupuncture

The procedure for an acupuncture operation is as follows. The needle is held close to its point with the left hand between the tip of the middle finger and the nail of the index finger supported by the thumb. While grasped in this manner, the needle is placed on the location to be punctured, which earlier was selected and distinguished by touch from the nerves. Next, one or two blows from the hammer held in the right hand drive the needle through the outer skin, which is a bit tougher. Then the hammer is laid down, and by turning the handle of the needle between the tips of the first fingers, the point is driven into the depth prescribed by the rules of the art (usually one half-inch, rarely an inch or more). In short, the needle is driven in until it reaches the area where the cause of the pain is concealed. When this has been reached, the implanted needle is held for one or two breaths, and then, when the needle has been withdrawn, the location is pressed with a finger; the intention is to force the spirit out by this route.[6] Needles of the second type are inserted merely by twisting them between the tips of the thumb and the middle finger. Those with a sufficiently practiced hand drive the needle in by a blow from the index finger placed over the middle finger. In this way these experts first perforate the skin and then twist the needle. Still others employ for this task the small tube, which I mentioned, in order to prevent the needle from being driven in too deeply by an overly forceful blow.

The precepts of the art of acupuncture are numerous and are chiefly concerned with the hidden location of spirits, insofar as the subject admits a methodology. Consequently, when undertaking an operation, a physician must methodically determine by touch the proper location and depth of a puncture. Acupuncture and moxibustion are remedies for the same ailments, and in general, locations that may be burned may also be punctured. We will point out these locations in the report on moxibustion. The common people, relying on experience alone, also dare to employ the needle without the advice of an expert Tensasj. The only caution they observe is to avoid piercing nerves, tendons, and more prominent arteries. Having made these preliminary observations, I will complete my subject, the cure of colic by acupuncture, with a few remarks.

The Japanese employ anticolic acupuncture on the region of the epigastrium.[7] This requires in all nine punctures arranged in three rows in the shape of a parallelogram with a distance of two inches (for male adults) between punctures [Plate 6]. These three rows are distinguished from one another in acupuncture instruction by having their own names and rules. The first row, called Sjoquan, is made just below the ribs; the second row, named Tsjuquan, occupies the midpoint between the navel and the mucronate cartilage; the third row, termed Gecquan, is located one half-inch above the navel. I have a number of times witnessed the employment of this method of acupuncture, which is illustrated for the reader in this chapter. Usually, after the three rows of punctures had been made according to the instructions of an expert and to the proper depth, the colic pains, called Senki, ceased immediately as if by magic.[8]

The Japanese also combat colic with their other method, moxibustion, but are not as successful in driving it out. Cauteries are burned on both sides of the navel at a distance of two inches from it. The two points are called Tensu, a term that is famous because such burning is common, and a term that is known even to persons who have never experienced the rites of this surgery. Moxibustion, however, will be discussed in its own chapter.[9]

The common people have recently become acquainted with an especially effective anticolic medicine that is today very frequently employed against colic and particularly against cholera, which is quite common and fatal in this region. The medicine is also very frequently employed: against chronic abdominal pains, called Saku, which are related to colic and are equally endemic; against other abdominal pains that are rooted in the intestines beyond the reach of a needle and do not readily yield to either fire or metal; and against other diseases that I shall not list. The medicine is a powder called Dsjosei in the speech of the common people and Wadususan in the language of the learned. It is only sold in the village of Menoki in the province of Oomi[10] under the seal of its inventor, who obtained a monopolistic privilege by a religious fraud. He told the following story. Jakusi, the Japanese Apollo-like god of medicine,[11] had revealed to him in a dream that the herbs growing here and there in the vicinity and on a mountain very famous in myths were a remedy for the common abdominal ailment. Successful results brought fame to the medicine, credence for its inventor, and wealth for his entire family. Having become rich quickly, the family was able to repay the god for its good fortune by building three temples with statues of the god who had generously revealed the secret. These temples are located in the area of the three shops where the powder is prepared with the aid of mills. I brought a supply of the powder with me from Japan. It is more bitter than bile, but German stomachs derive no more relief from it than from our own medicines. The recipe is kept secret by the family. However, I noticed at a shop that a bitter species of costus,[12] called Putsjuk, provides the basic ingredient. The Dutch export large quantities of Putsjuk from Surat[13] to Japan. Putsjuk has a variety of virtues and is very commonly employed in Japan, where, except for the root of the *sisarum montanum coraeense* (*ninsin cleyeri*),[14] no exotic vegetable is in greater demand.

Notes

1. Sakki (sake): an alcoholic beverage made from fermented rice.

2. gymnosophist: Kaempfer uses the term as a synonym for a Brahmin, a member of the highest caste, that of the priests among the Hindus. Members of this sect of Hindu philosphers wore little or no clothing, observed ascetic habits, and devoted themselves to mystical contemplation.

3. fire: moxibustion.

4. metal: acupuncture. Acupuncture, a principal therapeutic technique of Chinese traditional medicine, was fully developed before 2,500 B.C. and became popular in Japan during the sixth and seventh centuries A.D. In oriental thought, both the universe and humanity are governed by the forces of Yin and Yang. Yin is dark, passive, and female, while Yang is light, active, and male. The proper proportion of Yin and Yang maintains a state of well-being in nature and in humans. The human body is composed of five elements: wood, fire, earth, metal, and water. Other groups of five are also involved, for example, the five astrological forces and the planets Jupiter, Mars, Saturn, Venus, and Mercury. The forces of Yin and Yang operate in the body through twelve meridians, or channels, along which are located as many as 1,000 points that are said to have direct connections to specific organs (liver, kidney, stomach, etc.). Charts and models identify the individual points related to each organ. By the insertion of thin needles (acupuncture) or by the burning of cones of dried *Artemisia vulgaris* (moxibustion) the flow of Yin and Yang can be adjusted and thus health maintained or restored. In contemporary China and Japan acupuncture is still frequently employed, sometimes as an anesthetic in connection with Western medical procedures. The needles may also carry electrical stimulation. Wilhem ten Rhyne (1649–1700), a Dutchman who served as the physician on the island of Deshima, Nagasaki from 1674 to 76, composed the Western world's first detailed treatise on acupuncture, *Dissertatio de arthritide: mantissa schematica: de acupunctura: et orationes tres I. de chymiae et botaniae antiquitate et dignitate. II. de physionomia. III. de monstris* (London, 1683). See Robert W. Carrubba and John Z. Bowers, "The Western World's First Detailed Treatise on Acupuncture: Willem Ten Rhijne's *De Acupunctura*," *Journal of the History of Medicine and Allied Sciences* 29 (1974): 371–98. Kaempfer followed with his report on acupuncture as a cure for colic.

5. *Artemisia*: a genus of plants with a markedly bitter or aromatic taste. Wormwood, mugwort, and southernwood belong to this genus, which was apparently named in ancient Greece after the goddess Artemis (Diana). Kaempfer associates the name with the vassal ruler of Caria, who built (or at

117

least started construction of) the celebrated Mausoleum for her deceased husband (and brother) and then apparently died of grief over his death.

6. Schooled in medical traditions of Greece and Rome, Kaempfer naturally assumed that the insertion of the needle was intended to facilitate the exit of harmful vapors or winds.

7. epigastrium: the upper-middle part of the abdomen, including the area over and in front of the stomach.

8. Wilhem ten Rhyne concludes his *De Acupunctura* with an account of a similar procedure:

My guide for the journey to court, a garrison soldier of the Emperor of Japan, had emerged from a holocaust and, being exceedingly hot, he drank enough cold water to quench his thirst. A terrible pain, but one which did not radiate to his flanks, seized his stomach. In addition, from eating and drinking to excess as well as from being unaccustomed to the sea, he remained ill for a number of days with frequent nausea and vomiting. At first he attempted to cure these ailments with warm Japanese wine and ginger, but this did not relieve the pain. He blamed the persistent trapped wind for which he resorted to acupuncture. In my presence he himself performed the acupuncture in the following manner (from this case, reader, form your judgment about the others). Lying on his back, he drove the needle into the left side of his abdomen above the pylorus at four different locations. (For this task, he cautiously held the point of the needle with the tips of his fingers.) While he tapped the needle with a hammer (since his skin was rather tough), he held his breath. When the needle had been driven in about the width of a finger, he rotated its twisting-handle. He pressed the location punctured by the needle with his fingers. No blood, however, appeared after the extraction of the needle: only a very slight puncture mark remained. Relieved of the pain and cured by this procedure, he regained his health.

9. Moxibustion is the subject of Observation XII.

10. The province of Oomi included Lake Biwa near Kyoto. For Kaempfer's visit to the village of Menoki, see *History of Japan* 3:30-33.

11. Jakusi (Yakushi) is the Japanese Healing Buddha. The temple at Menoki included a gilded statue of the god, whose head was surrounded by a crown of rays and in whose left hand was a scepter.

12. Costus root comes from the Kashmirian plant, *Saussurea lappa.*

13. Surat: a city and port in western India on the Gulf of Cambay. During the sixteenth and seventeenth centuries, Surat was one of India's most populous

and active trade centers where both the British and Dutch established factories.

14. *sisarum*: the parsnip.

Observation XII
Moxa:
An Excellent Cautery Much Employed by the Chinese and Japanese

I
Introduction

There are three Helicons in Asia as far as it extends to the extreme east and the opposite side of the world.[1] All the sciences and arts known to the people of the vast Orient are derived from the three Helicons of the Arabs, Brahmins, and Chinese. The Orientals do indeed have different climates, languages, customs, and religions; in addition, they cherish quite different principles, precepts, materials, and procedures for medicine and healing (I will not discuss other fields) according to the Helicon from which they have drawn their knowledge. Nevertheless, when the question of the causes of diseases is raised, all these people instantly, and as if they were a chorus, place the blame upon winds and vapors. They appear to assign almost every pain and condition, as does Hippocrates (*On Winds*),[2] to winds. And they claim that these ailments are best removed by burning cauteries or, to use our terminology, actual fire. Do I mean fire or the fire of glowing iron? Orientals reject as an inane and futile savagery the practice of employing the combined forces of Vulcan and Mars on the human body. They further consider such a practice truly unworthy of a rational physician whose sole consideration when burning ought to be how he can resolve the painful impacted matter or how he can drive out the resolved wind and lead it into the open air. Consequently, they prefer to apply gently glowing and slowly burning fire—in a word, fire of a kind that gradually and safely draws out the imprisoned cause of disease by the power of aperient salt. They reject the kind of fire that savagely attacks and eats away the skin and flesh through the force of astringent and corroding vitriol. According to the same principles, the ancient Arabian, Egyptian, and Greek physicians, from whom we Europeans learned the art of healing, chose burning mushrooms and fiery roots of soapwort and birthwort instead of glowing iron. Some of these ancient physicians favored the use of hot sulphur and box-wood spindles dipped in boiling oil for controlling body ailments. Readers who desire information about the variety of caustic materials and methods of burning may consult L. Mercado[3] and M. A. Sever-ino,[4] among the more recent authors. In this chapter, I will only discuss the Asian cauteries employed in our time.

II
Various Materials for Burning in Asia

I have observed that the Arabs, as well as the Persians and the Mongul Mohammedans who are imbued with the Arabian sciences, are satisfied to employ only one material for burning: cotton cloth dyed with Isatis, or woad; the French call it cotton bleu.[5] They roll this cloth as tightly as possible in the shape of a cylinder with a diameter of one and one-half inches and a length of two inches. The cylinder is placed on the location to be burned, and when the top has been set on fire, it is allowed to burn and to be consumed gradually until the entire cylinder has been reduced to ashes. This burning process requires a long and almost intolerable period of time; it lasts for one quarter of an hour, sometimes longer, before the intense heat dies down. The flesh underneath is eaten away to such a depth that sometimes an incurable ulcer results. A number of such foul and wretched cases were brought to me for treatment in these countries.[6]

Once the burning has been completed, the surgeon's task is limited to anointing the wound daily and to promoting suppuration after the scab comes off. I believe that the excruciating pain and prolonged inconvenience are the reasons why the people in fact employ these cauteries with less frequency than is prescribed and inculcated in schools and books. I mentioned glastum, or woad. The Arabs, con-vinced that woad contributes significantly to the effectiveness of the fire, insist that the material for burning cauteries be dipped in its liquid. This belief, they claim, is not an empty one, but one based on many centuries of human experience. If we accept the assertion of the European common people, the application, beneath the nostrils, of a smoking linen cloth that has been dipped in woad more effectively dispels paroxysms suffered by epileptics (so-called demoniacs) than smoke from white linen or any other substance. Certainly, as a surgeon in the Indies, I have discovered the advantages of employing cloths dyed with Isatis as against other cloths for fomentation (a term used in surgery)[7] and bandaging of inflammations.

The Brahmins, whom the Greeks call gymnosophists,[8] are the sages, theologians, and physicians of the heathen Indians. The Brahmins, as well as the other heathens who learn from them, prepare cauteries not of a single type but of a variety of types in accordance with the variety of ailments. They claim that the causes of diseases are concealed and that the nature of each and every disease is different. Consequently, a single type of cautery cannot possibly be appropriate for the great variety of ailments. Experience, they claim, has taught them which cautery should be selected as fit and proper to the nature of a given disease. As a foreigner, I did not have the opportunity to examine the variety of cauteries. Sooner will a person separate Hercules from his club than learn the secrets of their arts from these jealous doctors. The most common of their cauteries (less use is made of other cauteries) is the pith of the *Juncus* that grows locally in the marshes.[9] It makes no difference which variety is employed, provided that it is thicker than the *Scirpus*.[10] The pith of the Juncus is dipped lightly in oil of the sesame plant, which the fields produce in abundance. The burning of human skin is then undertaken in the usual fashion. I have seen the Malays, the Javanese, and the Siamese utilize this very same pith for cremating their dead. Unless I am mistaken, other neighboring nations do the same thing.

Advancing beyond the Ganges, I find that the Chinese and Japanese employ the very best and most commonly utilized material for cauteries. With this material, they claim, cauterization was practiced from very ancient times before the discovery of medicine itself or any knowledge of surgery. Its use, they say, at last reached them through a tradition extending over many centuries. This native flax, which is universally called moxa, enjoys the greatest repute among the Chinese for its powers and antiquity. All the nations (Japan, Korea, Quinam, Philippine Islands, Formosa, Cochin, and Tonkin) that are trained in Chinese literature and learning have from centuries past adopted this flax for cauteries and use the term moxa for it. I here propose to present the history of moxa. I rely on the reader's kindness to pardon my use of Japanese terms as substitutes for the Chinese terms that are perhaps more desirable. Because of my greater knowledge of and contact with the Japanese, their terms are easier and more familiar to me than those of the Chinese.

III
Preparation of Moxa

Moxa is soft down or tender flax. It is ash-colored, very readily takes fire, burns moderately, has scarcely visible sparks and a moderate heat, and burns slowly until it is entirely consumed to ashes. Moxa is made from thoroughly dried leaves of the young *Artemisia vulgaris* (*latifolia*). The leaves are hung and aged in the open air for a long period. Artemisia, which will yield moxa, is not collected at random, but only on those days when, according to astrological calculations, the favor of heaven has endowed the artemisia as greatly as possible with the power of the stars. The assigned time is the first five days of the fifth Japanese month, called Gonguatz gonitz;[11] according to the Gregorian calendar, these days occur in June and occasionally in the end of May. The Japanese begin their year with the new moon closest to the midpoint between winter and the vernal equinox. The artemisia must be picked early in the morning when it is dripping with the dew of night. Then it is hung in the open air on the western side of the house until sufficiently dried out. Next, it is stored beneath the roof. The longer the artemisia ages, the greater the potency and tenderness that can be expected from its down. As a result, many people take pains to preserve the plant for as long as ten years. The Japanese call the young artemisia Futz; they call the mature adult plant Jamoggi.[12] As I noted elsewhere, it is customary in both China and Japan for a man to change his name both when he advances in age and achieves a new dignity. According to the same custom, plant names (not to mention the names of other things) change very often according to the degree of perfection and diversity of use. This practice lends distinction and luster to things, but it very seriously taxes one's memory. The preparation of moxa requires no special skill. First, the leaves are vigorously ground in a pestle to the softness of coarse flax. Next, they are assiduously rubbed and worked with both hands so that the harder fibers and membranous chaff are removed. What remains is the much desired, extraordinarily tender, homogeneous, and very pure down with which nature has devotedly endowed the royally named and delicate artemisia in preference to other plants.

IV
Persons and Illnesses

The burning of this down is in no respect terrifying for a human being. No sight of a blazing substance confronts the eye; rather, the pleasing scent of mildy glowing flax greets the nose. No significant pain troubles the patient, as happens with other cauteries, except perhaps from the cones that the Japanese call Kawa kiri or skin-cutters. These are the first three cones that are burned successively on a single location. The obedient and submissive people employ the term Kawa kiri metaphorically for burdens imposed on them by new rulers— regulations that are first hard to endure but are scarcely troublesome when one has become accustomed to them. I have countless times observed mere boys submit to being burned on various parts of the body without a sign of pain. Indeed, the Japanese have no reservations about burning children or old people, rich or poor, male or female; they spare only pregnant women who are not thoroughly accustomed to being burned. Burning with artemisia is undertaken either as a protection against or a cure for diseases. Physicians recommend burn- ing chiefly to preserve good health; it is the healthy more than the sick whom they order to be burned. A cautery, they assert, is both a healing substance for a present illness as well as a preventive against future illness. Consequently, in the extreme Orient, those who look after their health take care to be burned every six months. This practice is observed so scrupulously that even prisoners condemned for life are permitted to come out and enjoy the benefits of moxibustion. Preventive burning requires a few tiny cones on a single location; curative burning requires more and somewhat larger cones, if the vapors are to be brought from their hidden depths.

If you ask for what illnesses burning is appropriate, the Chinese as well as the Japanese will reply, Burning is appropriate for all the various diseases in which the imprisoned vapor causes a dissolution of solids, pain, and an impairment of proper functioning. The result is that moxibustion is prescribed for almost the entire list of illnesses. With its heat moxibustion burns away illnesses in a short while, or so they believe on the basis of daily successes. The dark people of the tropics, who learned the technique from their neighbors, also proclaim the power of moxa. They began to employ burning long after those who discovered it. When attacking bodily ailments, they use much larger cones as is demanded by the stubbornness of the condition and the tightness of the imprisoned matter. In recent years, the Dutch in the Indies have discovered the effectiveness of moxa against arthritis and its offspring, gout and sciatica.[13] The fire breaks up the particles of the tartar of Rhine wine[14] that assail the periosteum,[15] and it dissipates the sharp fluid lodged about the cavities of the bones in cases of sciatica and the other arthritic discharges. However, the cone must be larger and it must be applied soon enough before the fluid tears apart the membranes and muscles that have been overly stretched by the vapor. When such a tearing occurs, the lacerated capillary vessels flood and fill the openings with humors and produce serious abscesses that cannot be dissolved except by cutting. It must be noted, however, that while moxibustion succeeds in a warm climate, it is less effective in our colder climate. In a warm climate, the body is more permeable, the matter more fluid, the pores more open, and the muscles and membranes more relaxed.

We must further note that by moxibustion pain is in fact checked but by no means ended. Pain is checked in the area where the burning cone drew out harmful particles or seriously burned the periosteum and rendered it insensitive, but pain subsequently makes its way to other locations. The Brahmins promise that pain will never recur, if their instructions are followed after burning: all drink produced by fermentation (such as wine, beer, and intoxicating spirits) as well as all types of meat must be excluded completely from the patient's diet. These items, they claim, supply new crudities that as the blood circulates, glide down to the legs and lodge once again in the periosteum.

Reverend Busschof, a pastor at Batavia in the Indies,[16] went too far in recommending moxa to his European countrymen as an infallible remedy for the gout.[17] Indeed, I am afraid that his recommendation has disappointed the expectations of many Germans. Recently, and quite justifiably, the renowned professor at Giessen, Dr. Valentini,[18] Adjunct of the Academy of Sciences in Germany, complained of this in a printed letter to Dr. Cleyer.[19] That very learned—as are all things from this very distinguished gentleman—letter was delivered to Dr. Cleyer in my presence. In cases of epilepsy and all chronic head ailments, their dark neighbors are more inclined than the Chinese to employ moxa as a remedy. They burn the entire sutura coronalis[20] with a long and large cylinder; sometimes health is restored in cases considered hopeless by physicians.

V
Locations for Burning

The practitioners of this pyrosophic surgery disagree considerably over what locations ought to be burned in order to prevent or to cure particular ailments. Often the disagreement rests on superstition and vanity! When each person, relying, as he claims, on the experience of himself or his teacher, has indicated one location or another, and these individual opinions have been collated, virtually every part of the body is liable to burning, often for a single condition. The common people, however, rarely stray from the accepted locations that have been handed down from antiquity and that they know from printed charts in their possession. Vanity is an even greater factor in the choice of the time at which, they believe, the stars permit or forbid the burning of this or that part of the body. The practitioners reject burning on each day and hour when the spirit of the stars impregnates (so the Japanese theorize) a part of the body that otherwise may be burned with success for a given condition. Consequently, if you observe their collective prohibitions, there will rarely be a proper time for burning.

The paramount consideration is the selection of locations for burning that are most advantageous either for drawing out vapors or for removing the harmful matter from the affected part. The practitioners of moxibustion claim that they have precise knowledge of these locations as handed down by their teachers, while they also boast of their own experience. I have seen no part of the body more scarred from moxibustion than the back: both sides close to the spine and as far down as the loins. The backs of the Orientals and especially of the Japanese (male and female) bear so many marks and deep scars or ulcers that you would swear they were brutally whipped. In the eyes of the Orientals, these disfigurations do not at all detract from a person's beauty. It is an easy matter, especially for the Japanese, to bare the upper portion of the body, and this is their practice even when undertaking slight tasks. The Japanese, a people not given to underwear, lower their robes, which are gathered far down at the loins, in order to prevent soiling by perspiration. Hence, scars on both sexes come into public view.

VI
Procedure for Burning

No great skill is required for the execution of moxibustion, which is performed in the following manner. A small amount of moxa is rolled with the tips of the fingers into a cone. Ordinarily, the height of the cone is nearly one inch and the diameter at the base is somewhat less. The cone is placed on the location to be burned. If one wishes, saliva may be applied under the base of the cone for adhesive purposes. Next, a spark is applied to the apex with a very slender, glowing rod or stick, which the Japanese call Senko. A tiny flame burns away the cone in a short time. One or more additional cones may be placed on the same location and burned. The person supervising or performing the operation determines how many cones are sufficient for his purpose or are prescribed by the rules of the art. Surgeons of moxibustion are called Tensasj, touchers, or literally, penetrators by touch, because they use their fingers to find the locations for burning.

The rods are the same ones that the heathen monks burn when worshipping their idols and marking off the devotional hours in the temples. In the same way, we measure the periods of guard duty in military camps with flax. The rods burn sparingly and are consumed slowly, but they emit a very fragrant scent. The rods are made from the extremely slimy bark of the Taab or Taab no ki tree, which is the *Laurus Japonica sylvestris*,[21] the largest tree in the area. Bark is mixed with aloeswood[22] or its most precious resinous part, called calambac, and various other fragrances, according to the fancy of the individual; all these ingredients are in the form of powder. The mixture is reduced to a thick pulp by adding water. After long and adequate kneading by hand, the pulp is placed in a vessel with many round holes at the bottom and is pressed down by a heavy weight on top. Rods thinner than an oaten stem come through the openings below. They are immediately placed on lathes and thoroughly dried in the shade. The shops stock these rods for use as smoking candles. Bundles of rods wrapped in paper are sold for purposes of moxibustion. The truth is that these rods are unnecessary additions to surgery by moxibustion; they contribute merely elegance to the operation and nothing else. The practice of the common people is quite adequate: they ignite the cone with a burning splinter or any little twig.

The essence of the art lies in the knowledge of the proper location to be burned for a given malady. A European is likely to judge as most appropriate for withdrawing vapors (since this is the sole purpose in burning) a location closest to the afflicted part. For the most part, however, the practitioner selects a remote location, often one not related to the troubled part by any known anatomical connections except through the common integument of the whole body. For just as a certain Lithuanian nobleman considered it absurd that an enema was prescribed when his head ached, so, too, do foreigners think that instant successes by moxibustion are illusory when cauteries are applied to a member and a location that are remote and completely free from pain. The following are a few examples. The shoulders are burned with success for indigestion and loss of appetite; the spinal vertebrae are burned to relieve the pain of pleurisy; the adducting muscle of the thumb on the same side as the pain is burned to alleviate toothache. What anatomist, no matter how skillful, could demonstrate any particular relationship between these parts of the body?

VII
Rules

The rules and requirements of moxibustion are numerous with respect to location on the body and time for burning as well as number of cones, position of the body, diet of the patient, and other circumstances. Among the foremost general precepts are the following. As far as possible, tendons, arteries, and veins must be avoided; for caution's sake, the practitioner must examine both by sight and touch before burning. Once a patient has assumed either the sitting or the standing position for the determination of the location to be burned, he must maintain this position during the operation. The patient should sit on the ground with his legs crossed and hold one hand to each cheek, since this position reveals the muscles and the interstices and corresponds most to the natural position of the fetus. If a patient is to be burned on the legs, he should sit on a chair with his feet hanging down immersed in warm water. In this way, exhalation is artificially promoted in the parts of the body more removed from the source of heat. Patients with a weak constitution should be burned with no more than three cones on any location. Patients with a stronger constitution may be burned with ten, twenty, or more cones depending on the nature of the malady. There is no rule of the art that answers the question whether a number of cones may be burned on different locations simultaneously or alternately. The decision depends upon the tolerance of the patient and the leisure of the practitioner. The practitioner should examine the burn on the day following the operation and for several days thereafter. If the burn has not suppurated and looks dry, this is considered a bad sign indicating a weak constitution. In this event, the practitioner should promote suppuration by applying crushed onions. The previous observations were gathered from occasional conversations with surgeons of the pyrotechnic art.

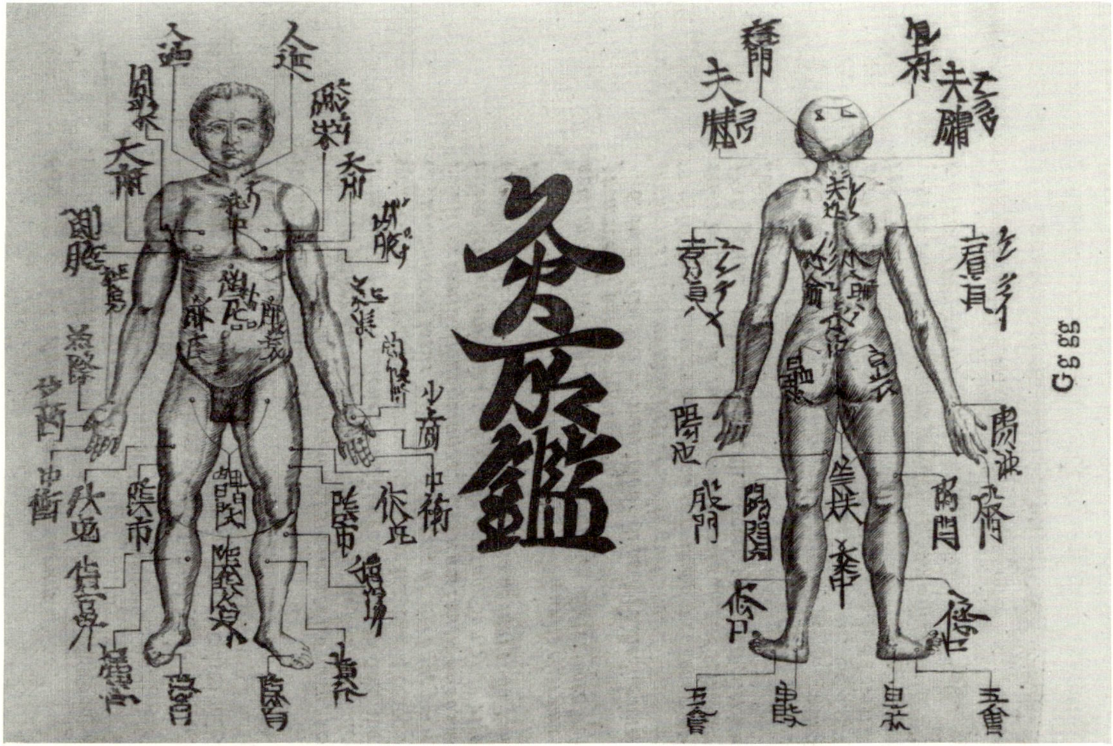

Plate 7. Moxibustion Points

Charts printed in Chinese and Japanese present particular rules for burning. I have here included one such chart, which I translated as well as I could, considering the nature of the verse and Chinese philosophy. I have also included two figures on which the locations to be burned for particular diseases are shown and their proper names indicated. These are sold in bookshops and by quacks who hawk them in the streets and marketplaces. The sellers invite people ignorant of this type of surgery to buy an inexpensive chart that will teach them in summary form the rules of the prolix art. The translated text is printed below; my own comments are enclosed within brackets.[23]

The method of burning is presented through certain propositions (in verse). From these propositions anyone can clearly understand the whole art of burning.

Kju sju Kagami,
or
Illustration of the Locations to Be Burned

Section I

1. For headache, vertigo, fainting fits, Dseoki, and blurred vision occasioned by frequent attacks of Dseoki; for shoulder pains following a headache; for asthma and shortness of breath; it is helpful to burn that part of the human body which is called Koko. (Dseoki is an inflammation of the face, a native ailment occasioned by scurvy. Victims of Dseoki suffer from a swelling and fever of the face and often of the entire head; this condition is induced by trivial causes such as bathing, excessive drinking and exercise. Dseoki is often followed by an inflammation of the eyes.)

2. For childhood diseases, especially swelling of the stomach, diarrhea, loss of appetite, itching and abrasion of the nose; as well as for night blindness in adults: burn the region of Sjuitz, or both sides of the Eleventh (sc. vertebra), with fifteen or sixteen cones. Keep a space of one and one-half suns (about two or three inches or finger widths) between these two places which must be burned.

(Note 1. The region of Sjuitz, or Eleventh, takes its name from the eleventh vertebra, on either side of which one must burn. The count begins with the fourth vertebra of the neck, which, to be sure, protrudes more than the others when the head is bent forward toward the chest. All subsequent numbering of the vertebrae is likewise based on taking the fourth vertebra as the first.

Note 2. Sun is a measure of length used in Japan. One kind of sun is longer and is used by merchants. The other kind of sun is shorter and is used by builders and artisans. But neither type of sun applies to the process of burning. For purposes of burning, one sun equals the length from the second joint to the tip of the middle finger of the particular person on whom the operation is to be performed. This measurement best corresponds proportionately to the measurements of the rest of the body.)

3. For Sakf (chronic and intermittent pain of the abdomen); for Senki (painful endemic colic); for Subakf (intestinal gripes caused by worms): burn on both sides of the navel at a distance of two suns. This location for burning is called Tensu.

4. For both menstrual obstruction as well as for excessive flow; likewise for white flux; for hemorrhoids and ulceration of hemorrhoids; for Tekagami (a painful cold or chronic and intermittent catarrh): burn the location Kisoo no Kitz on both sides with five cones made of artemisia. To find this location, go straight down from the navel for four suns. Then go right and left for four suns. The points for burning are separated by eight suns.

5. For difficult childbirth: burn three cones on the very tip of the little toe of the left foot. This instant relief promotes delivery.

6. For a lack of milk in wet nurses, five cones must be burned on the middle of the chest between the breasts.

7. For arthritis and sciatica; for pains of the legs; for strangury:[24] about eleven cones must be burned on the thighs at a distance of three suns above the knees (this is the location for artificial issues).

8. For swollen and painful stomach; for heart pain arising from daily fever; for pains of the belly and loss of appetite: burn six cones above the navel. The location for burning is four suns directly above the navel.

9. For pains of the hips and knees; for lassitude and weakness of the legs; indeed, for any languor of the limbs and of the whole body: burn Fusi. (Fusi is the location on either thigh that the tip of the middle finger reaches when the arms and hands hang straight down in a natural position.)

10. Persons who have hardness of the hypochondrium,[25] as well as persons who are subject to cold shivers and recurrent wasting fevers, will take care to be burned on the location Seomon. (Seomon is the location just beneath the last false rib on each side; burning on this location causes almost unbearable pain. I would have written Schomon or Sjomon, had I not heard the Japanese pronounce the "e" very short.)

11. A person treating gonorrhea will burn artemisia in the middle location Jokomon. (Jokomon is the location just above the pubic hair and is midway between the navel and the genitals.)

12. A person suffering from a catarrh, nasal hemorrhage, or vertigo, will obtain help by burning the location Fuumon with fifty to one hundred cones (successively on the same place). (Fuumon is the region of the sacrum.)[26]

13. A person who has growths from the anus or painful condylomata[27] must burn one cone at a distance of three suns from the extremity of the coccyx.[28] (The pain from burning is very severe!)

14. For prolapse of the anus, burn the coccyx itself.

Section II

1. Nindsin (the spirit of the stars) inhabits the region of the ninth vertebra in spring; Nindsin also dwells in the region of the fifth vertebra in summer, in the region of the third vertebra in autumn, and in the region of the fourteenth vertebra and both hips in winter. Burning must not be undertaken in these locations when Nindsin is present.

2. When the four seasons of the year are changing, no burning may be performed either at the location Seomon or in the region of the fourteenth vertebra. Burning undertaken at these times, instead of being beneficial, is particularly harmful because it stimulates or aug-

ments disease.

3. During rainy, damp, or hot weather, as well as on a cold day, there must be total and complete abstention from burning.

4. For three days before and seven days after burning, sexual intercourse must be avoided.

5. A gravely angered person should not be burned until his spirit has calmed. A fatigued person or a person who has just completed hard work must not be burned for a while. Let the same rule apply to a hungry person and to a person who has eaten excessively.

6. Before burning, it is bad to drink sake (a rich, alcoholic native drink brewed from rice); after burning, it is both good and healthful, because sake promotes the circulation of the spirits and blood. (The Japanese have known from antiquity that the humors circulate; however, they do not know where and how).

7. A person must fully comprehend and diligently observe the rule that within three days after burning he must bathe with sweet water. (The Japanese make daily use of such bathing and particularly of what is called a steam bath. By this means, I believe, they drive off venereal disease, which would otherwise destroy the entire nation.)

8. Drugs are designed to cure disease; burning is designed to prevent disease. Hence, even a person in perfect health should allow his body to be burned during the second month (March) and the eighth month (September). (Days for burning that enjoy the favorable influence of the stars are noted on the calendars.)

9. Before burning, a person should check the pulse. If the pulse is unduly rapid, one must proceed with caution, since the patient is suffering from a cold.

10. A person wishing to burn will mark the locations for burning after he has measured by saku and sun. The length of a sun is gotten from the middle joint of the middle finger; on the left hand for men, on the right hand for women.

Section III

Any woman who wishes to make herself immune to conception will burn three cones of artemisia on the navel.

Section IV

Any woman who desires children and wishes to prevent infertility must burn eleven cones on each side of the twenty-first vertebra.

Notes

1. Helicon: a mountain in Boeotia (a province of ancient Greece) sacred to the Muses and a source of poetic inspiration. Kaempfer here employs the term symbolically to stand for a creative tradition in science and art.

2. *On Winds*, in which diseases are regarded as caused by winds (*pneumata*), is one of the works comprising the *Corpus Hippocraticum*. See also Observation VIII, note 9.

3. Luis Mercado (1520–1606), *Praxis medica, nunc primum hic in lucem edita: in quatuor partes divisa* (Venice, 1608). Kaempfer refers the reader to "De febrium essentia" (4.1.162).

4. The celebrated Italian anatomist and surgeon, Marcus Aurelius Severino (1580–1656), *De efficaci medicina libri III* (Frankfurt, 1646). See also Observation VIII, note 7.

5. woad: a blue dye prepared from the leaves of the *Isatis tinctoria*, a cruciferous plant.

6. Between 1683 and 1689, Kaempfer had visited Persia, Arabia, India, and Ceylon.

7. fomentation: a hot and wet application for the relief of pain or inflammation.

8. Kaempfer also describes the Brahmins as gymnosophists; see Observation IX, note 2.

9. *Juncus*: a genus of the rush family.

10. *Scirpus*: an extensive genus of hardy plants that includes the bulrush.

11. Gonguatz gonitz: *gongatsu gonichi*, literally five months, five days.

12. Jamoggi: *yomogi*, or mugwort.

13. sciatica: severe pain in the area of the hip and thigh, especially along the course of the sciatic nerve at the back of the thigh.

14. tartar: an acid compound in the juice of grapes and deposited on the sides of casks during the making of wine.

15. periosteum: the tough fibrous membrane surrounding a bone except at the ends which are covered with cartilage.

16. Hermann Busschof, *Het podagra metz gaders desself seker geneazinge* (Amsterdam, 1674). An English translation appeared in 1676.

17. Gout (podagra): a metabolic disease characterized by an excess of uric acid with swelling and severe pain of the joints, especially the big toe.

18. Michael Bernhard Valentini (1657–1729), *Historia moxae cum adjunctis meditationibus de podagra* (Leiden, 1686).

19. The German surgeon Andreas Cleyer served on Deshima in 1682 and 1685 but was expelled from Japan for smuggling. Cleyer published an early and important work on Oriental medicine, *Specimen medicinae Sinicae, sive Opuscula medica ad mentem Sinensium* (Frankfurt, 1682).

20. sutura coronalis: union of the parietal and frontal bones of the cranium.

21. *laurus Japonica sylvestris*: wild Japanese laurel tree with leaves that are leathery and evergreen.

22. aloeswood: the soft, resinous, and fragrant aloe; agalloch.

23. Kaempfer's comments will be enclosed within parentheses.

24. strangury: painful and interrupted urination.

25. hypochondrium: the part of the abdomen beneath the lower ribs on each side of the epigastrium.

26. sacrum: the triangular bone made up of five fused vertebrae and situated at the lower end of the spinal column where it joins both hipbones.

27. condylomata: wartlike growths usually in the area of the anus or genital organs.

28. coccyx: small, triangular bone at the base of the spinal column.

Plate 8. *Tea Plant*

Observation XIII
The History of Japanese Tea

I
Botanical Description

It may appear that in undertaking a history of tea I am simply repeating the task of the distinguished Dr. Wilhem ten Rhyne, my predecessor at this station in Japan and my much honored friend. His full and quite accurate account was published by the distinguished Dr. Breyn in the appendix to his *Century of Exotic Plants*.[1] As it happened, however, a briefer sojourn in this kingdom and a somewhat more restricted way of life denied that most observant gentleman the opportunity to investigate more fully. He omitted certain facts worth knowing and relevant to a complete history. I consider these, as well as a brief repetition of the whole history, worth presenting here.

Tsjaa

Tea, a shrub with the leaf of a cherry tree, with the flower of a rose bush, and with fruit that is unicapsular, bicapsular, and usually tricapsular.

The tea plant grows slowly to a height of more than six feet. It has an irregularly branched and lignose root with a blackish exterior. Its stem is ramose from the bottom with very many irregularly extended branches and furculae. Its bark is delicate, sapless, thin, chestnut colored, whitish on the stem and a trifle green on the young shoots. The wood is somewhat hard and fibrous; the pith is very small and firmly adnate. Individual leaves, attached to very short carnose pedicles, irregularly occupy the branches with unfailing strength, unless they are plucked. They resemble the leaves of the *Cerasus hortensis fructu acido*[2] in substance, shape, color, and size. Young leaves, however, such as those picked for internal use, perhaps rather resemble those of the *Euonymus vulgaris granis rubris*,[3] color excepted.

In autumn one or two flowers come forth from the leaf axils. They

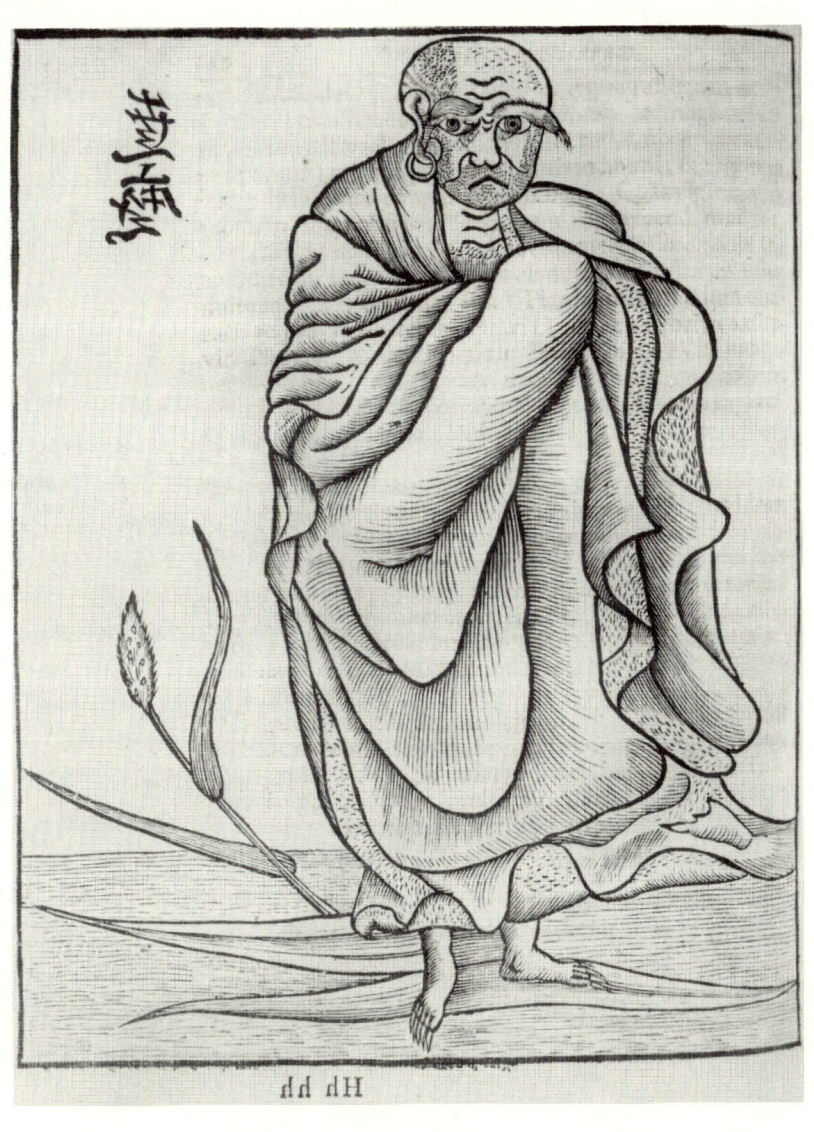

Plate 9. *Darma (Daruma)*

resemble wild roses, are an inch or more in diameter when spread (but not completely), have a faint fragrance, and are white and hexapetalous with circular, concave petals. The flowers stand on pedicles one half-inch long, initially thin but growing thicker, that end in little orbiculate squamae of an uncertain number (usually five or six), instead of the calyx. The flower is followed by plentiful fruits, which are unicapsular, bicapsular, or usually tricapsular like those of rice seed, a joint growth of three globular capsules, the size of a wild plum, united at the center on one common pedicle and separated by cushion-shaped depressions (where the fruits can be separated). A fruit consists of husk and nut enclosed tightly by its own capsule. The husk is green and inclines to black through ripening with an opaque exterior, a fat, membranous, and somewhat lignose substance. The outer surface develops openings as it dries (after a year on the plant) so that the nut is visible through individual cracks. The nut is nearly round, but a certain portion of it is compressed from resting against another capsule. It is covered with a thin, hardish, shiny, nut-brown shell very like that of a chestnut. The shell removed, there appears a kernel, reddish on the outside, of a firm and oily substance like the filbert. The taste is initially vapid and sweetish but soon bitter and extremely sour, like a cherry seed. The kernel inclines the taster to spit profusely, and having passed to the throat, it offends the gullet with a nauseous sensation that, however, quickly disappears.

II
Name

Tea, called Tsjaa by the Japanese and Theh by the Chinese, still has no character of its own accepted and approved by the universities. A character of its own is one that suggests the nature of a thing. Meanwhile, various substitutes have been used. Some characters at least indicate the sound of the word tea; others, the virtues and description of the plant. To the latter group belongs the character that represents the eyelids of Darma, a certain eminent holy man among the heathen [Plate 9]. The story behind it is both charming and relevant, for it indicates the date of the first use of tea.[4] With the reader's permission, I will briefly pursue this digression. Darma, the third son of the Indian King Kosjuwo, was a holy and devout man, a kind of pope of the religious. He was the twenty-eighth successor to the Holy See of Sjaka, also a dark Indian, who was born 1,028 years before Christ and founded Oriental paganism. In A.D. 519 Darma journeyed to China. He devoted all his energy to fill the people with the knowledge of God and a true—as he called it—faith that would lead them to happiness. But in his endeavor to serve mankind not only with doctrine but also with good example, he strove to win divine grace by living in the open air and variously chastising his body and mastering his passions. He ate only vegetables. His idea of the highest state of holiness was to pass his nights sleepless and alert in constant satori, that is, in contemplation of the supreme divinity.[5] To concentrate unceasingly upon God while denying the body rest and relaxation he considered the supreme achievement of a life of penance and the test of human perfection.

After vigils over several years, it happened that fatigued by long exposure and fasting, he succumbed to sleep. When he awoke, he cut off both his eyelids as the instrument of his sin and cast them in anger upon the ground in penance for his broken vow and as a precaution against the same thing happening in the future. Returning on the next day to the place of his punishment, he observed that by a wonderful transformation from each eyelid a shrub had grown—the tea plant. Either the plant had not previously existed or at the least mankind was ignorant of its virtue. As Darma ate the leaves (whether raw or prepared with water I do not know), he perceived a wondrous joy of spirit and strength to contemplate the divine. Because he commended this power hidden in tea leaves and the manner of eating them to his multitude of disciples, the use of this most noble plant readily grew and gained common approval as something about which no praise would be excessive. Meanwhile, the plant, as it lacked its own character, was conventionally represented by the eyelids of Darma. I am pleased to include here for the reader's inspection a drawing of that famous man, the type that enjoys great veneration among the heathen. Beneath Darma's feet is a reed on which, tradition relates, he traversed the sea and the rivers. So much for the name of the plant.

III
Supplementary Botanical Description

I gave a rather brief description of the plant to allow the reader to grasp more readily the idea. I will now add items that appear desirable for a fuller account. The stem often appears to have more branches at the bottom than it really has. When several seeds are deposited in a single hole, often several shrubs grow on the same spot, leading those less knowledgeable to believe that there is only one plant. In addition, older plants, which as is customary have been cut at the base of the stem, grow back denser and with very many branches and twigs, the product of a single root. The first year's shoots, whether growing from seed or from a stem that has been cut (as shown in the illustration [Plate 8]) are always fewer but much thicker and quite larger than subsequent ones. In time they become branched. The bark is firm and closely adnate; its cuticle is thin and occasionally falls off through dryness; the bast is green with an odor of hazel leaves, but somewhat heavy and unpleasant, and a taste that is a bit nauseating, bitter, wild and astringent. The wood is hard, composed of thick fibers, is green inclining to white, and while green has an unpleasant odor.

The branches are dense, not arranged in a pattern, slender, short, unequal, not marked with rings of annual growth, thickly set in no pattern with single leaves from whose axils protrudes a small, narrow bud. On fat, very short and green pedicles (whose back is smooth and front is somewhat compressed and carinate) stand leaves. The leaves are partially membranous and partially carnose with a maximum length of two inches and a width of one inch or less. Their shape, rather narrow initially, slowly broadens like a bow with a blunt tip. Often the shape is also oval, upturned lengthwise and irregularly undulated with a depression in the middle and the extremities a bit bent back. Each side is smooth with many hollows from depressed little veins and is shiny with a full and unpleasant green color quite softer on the back. The edge of the leaf is densely and unevenly serrated with little teeth somewhat upturned, rather hard and dull. A middle vein is depressed with a large hollow on the front but is remarkably prominent on the back. From both sides of it evenly branch out five, six, or seven thinner veins that curve back upon themselves before the edge of the leaf. Sometimes thinner stray little veins run between these. A fresh leaf has no smell and a taste milder than that of the bark, that is, partially wild and partially bitterish, astringent but not sour. Leaves vary greatly in substance, size, and shape depending upon their soil, situation, and age. Consequently, young leaves, such as the dried ones imported into Europe, give no true indication of the shape of full-grown leaves. The leaves have a certain malignant power and quality inimical to the brain, and by disturbing the animal spirits, they are able to intoxicate the mind and induce a trembling of the nerves. This effect is completely avoided by thoroughly drying the leaves so as to eliminate the narcotic and retain the pleasant virtue that refreshes the animal spirits.

The branches are adorned with flowers that continue to grow from autumn into late winter. Of the six petals on a flower, often one or two external petals, as if attacked by sphacelus,[6] do not grow as large as the others. The petals have an unpleasant and a trifle bitter taste that immediately seizes the base of the tongue. The hollow of the flower is filled with a dense array of white stamina that are, as in the case of the rose, exceedingly small with tiny, yellow heart-shaped tips. I counted 230 stamina on a single flower. The kernels of the fruit are very oily and quickly turn rancid so that when planted, scarcely two in ten germinate. The natives make no use of either the flowers or the kernels. I, however, have no doubt that the kernels possess an extraordinary virtue.

IV
Cultivation

We begin its culture with the planting of the seeds. The Japanese do not allocate individual gardens or fields for planting the seeds but merely the borders of the fields, regardless of the condition of the soil. The seeds are not planted in a continuous row so that the plants may grow into a hedge, but at moderate intervals so that the growing shrubs will not harm the fields with their shadows nor weave hindrances for those picking the leaves. At least six and not more than twelve seeds, still enclosed in their vessels, are placed in one hole one and one-half palms deep.[7] The large number of seeds is required because scarcely one in four or five germinates, as most seeds are either empty or are destroyed by rotting, a condition they very quickly contract. This marked tendency of the seeds to rot is the reason why, when brought to Europe and planted, they never germinate.[8] If one wishes to transport them safely for planting in Sicily, Spain, or Italy, I advise that they be placed in their native region in a pot filled with earth, so that the shoots that have grown from them may be dug out and planted at will. The same successful propagation does not occur with transported plants. While being carried through torrid India, they contract a wasting disease and are destined to complete their life span without vigor so that propagation and offspring are not to be expected of them. The growing shrub, if it enjoys an industrious owner, will be nourished with a mixture of horse manure and earth placed about it. Each year careful owners complete this operation that others neglect. When a plant is three years old and no longer young, it is permissable to pick the leaves, which by then are excellent and sufficiently numerous. At the age of seven a plant achieves the height of a man, but as it then grows rather sparingly and consequently produces fewer leaves, it should be picked thoroughly and cut down at the very bottom of the stem. In this way new and copious shoots grow up from the root, and by supplying a greater abundance of leaves, they repay the loss of the shrub with interest. Others defer cutting down a shrub until the tenth year.

V
Gathering of Leaves

At harvest time an owner of a large number of plants hires laborers skilled in this work. Since the leaves must be picked one by one and not torn off piecemeal, a servant could scarcely fill three catties[9] in a whole day, while a person engaged in this work from youth will pick nine or ten. The gathering is not undertaken on a single occasion, and it is usually completed by more than one effort. Those who have three harvests begin about the end of the month of Songuats, which is the first month of the Japanese year, beginning with the new moon preceding the vernal equinox either at the end of February or the beginning of March. This month, however, only furnishes a few leaves, and these, only two or three days old, are very tender and not yet fully unfolded. But they are of the very highest quality and on account of their price and rarity must be reserved for princely palaces and wealthy homes. Such tea is commonly called Imperial Tea or the Flower of Tea. (No one gathers the petals of the flowers or employs them as substitutes for the leaves, as our people wrongly thought. I believe that this error arose from the name "Flower of Tea" or was introduced by ignorant voyagers.) To this category of tea belongs Chinese Buu, that authentic and superb tea that is rightly dear even in its own native land. The second gathering (the first for others) begins in the second month, that is, about the end of March or the beginning of April. Both fully unfolded leaves as well as, if one wishes, half-unfolded leaves are picked. Before preparation they are to be classified by quality and size. Since partially unfolded leaves somewhat achieve the goodness of the previous type, whose label they consequently bear, they must be separated from the rest on the basis of excellence. The last (the second for others) and the most abundant gathering occurs in the third Japanese month when the unfolded leaves are plentiful and luxuriant. Some owners who have dispensed with prior gatherings complete their entire harvest of tea with the one gathering in this month. With precise distinction, the leaves are assigned to different classes of goodness: itziban, niban, and sanban, or first, second, and third. The third class contains leaves that are two months grown and coarser, for use by the common people.

VI
Different Types of Tea

Hence, there exists a classification of tea into three types. The first type contains leaves of a very tender age. When prepared, it is commonly called Ficki tsjaa, that is, ground tea, because it is reduced by grinding to a powder, which is sipped in hot water. It is also called Udsi tsjaa[10] and Tacke sacki tsjaa from the localities in which most of it is gathered. This tea is nobler than the others because it is picked from shrubs three years old (considered the best of all ages) and because of the good soil. Soil and age of a shrub chiefly contribute to goodness as well as to the growth and largeness of the leaves. Largeness is not always an indication of goodness, unless it is joined with exceptional tenderness, for largeness may merely indicate old age. Chinese Teh Buu corresponds, as I have said, to this type of tea. The second type contains older leaves. When prepared, it is called Too tsjaa, that is, Chinese Tea, because it is usually prepared in the Chinese fashion. In Japan this type of tea is usually divided by dealers and vendors into four classes differing in goodness and price.

The first class contains young leaves that are gathered at the very beginning of spring when each new branch bears not more than two or three leaves, either fully or partially unfolded. In Japan one pre-pared chin (foreigners call it a catty), that is, a Dutch pound and one quarter, costs (provided I, being a foreigner, was not misinformed) a sja mome and more. Foreigners would say one or more taels, or ten to twelve silver mace, that is, 70 to 84 Dutch stivers, since one mace equals seven stivers.[11] The second class contains leaves a little older that may be gathered in any month. One catty of these costs six or seven silver mace. The third class includes some coarser leaves. A catty of these sells for four or five mace. To this class belongs the majority of the tea that is shipped from China to Europe and today costs five, six, or seven guilders in Holland. The fourth class contains leaves that are gathered promiscuously and indiscriminately when each new branch is estimated to furnish ten or at the most fifteen leaves. One catty costs three silver mace, at which price it is sold daily in the streets by vendors. This, then, is the normal, everyday tea employed by the majority of the people.

Leaves left on a shrub can very quickly alter their classification according to size and goodness, so much so that should the gathering be neglected, they may change from a higher to an inferior category even in a single night. The third type of tea is called Ban tsjaa. It is a melange of leaves from the last gathering, which, to be sure, are rather stiff and unfit for preparation in the Chinese fashion (that is, for drying and curling). Hence, however prepared, they are destined for use by country folk and the plebeian multitude. These leaves are said to have a persistent strength that diminishes neither from extended exposure to the air nor from cooking. Tea belonging to higher categories reacts differently. Its power, though more subtle and noble, is highly volatile, enduring neither exposure to the air nor cooking without a marked loss.

Mention has been made of Udsi Tsjaa. Lest anything be omitted for this account, I will include a few remarks about this tea. Udsi is a small town situated toward the seacoast and not far from the capital city and ecclesiastic seat, Miaco, in the province of the same name. Its exceedingly benign climate so favors the cultivation of the shrub that tea produced there is considered superior to others. In this place is located the celebrated mountain, with the same name as the town, reserved for the cultivation of the tea that supplies the imperial residence.[12] The Chief Prefect of Tea at the Imperial Court, who is responsible for the care of this mountain, also supervises through his deputies the cultivation of the shrubs as well as the gathering and preparation of the leaves.

The mountain has a delightful appearance. A surrounding moat wards off animals and strangers. The arrangement of shrubs in rows gives the appearance of walks. The paths and shrubs are swept and cleaned each day, and the guards are charged with zealous caution lest the leaves be dirtied. For increased protection many shrubs are even girt with hedges. As the time for gathering approaches, the pickers abstain from eating fish and unclean food for two or three weeks in advance so as to avoid contaminating the leaves with an impure breath. During the gathering period itself, they wash their bodies two or three times daily either in the bath or in the river. In addition, they dare not touch the leaves with their bare hands. Gloves must be worn while picking. Leaves prepared according to a fixed art and wrapped in paper bags are placed in urns or murrhine jars[13] and protected with common tea tightly packed all around. The director of the garden immediately sends the leaves duly packaged and attended by many guards to the imperial palace as a mark of reverence for His Royal Majesty. Hence, it happens that this tea acquires a notorious reputation for the incredible costliness of its production and delivery. For when expenses are calculated, one chin or catty costs not less than thirty or forty sju mome or taels, that is, 42 or 56 ounces of silver. Nor does the supervisor, in rendering accounts to the treasury, blush at increasing the price of some tea to one oban (a gold coin worth a hundred ounces of silver) or sometimes even to a hundred taels, or 140 ounces. Occasionally, a small urn, containing not more than three or four catties, arrives at the palace with a retinue of nearly two hundred men. An attendant at the imperial court presented this tea to me with this assertion: "Drink heartily and with pleasure; one dish costs one itzibu." An itzibu is a square gold coin worth one and one quarter of our gold coins.

VII
Preparation

This is the manner of preparation. Freshly picked leaves are roasted on an iron plate. When roasted and still hot, they are rolled on a mat with the palm of the hand until they curl. Roasting not only dries the leaves but also removes the malignant element offensive to the brain so that the leaves are mellowed for human consumption. The reason for rolling the leaves into a compact or small mass is convenience of storage. This operation is accomplished in Tsjasi, public roasting-houses in the cities, so equipped that each person may bring his own leaves for roasting, since private persons rarely have either the knowl-edge or instruments for correctly preparing the leaves. The roasting-houses are equipped with:

1. numerous ovens (five, ten, or twenty) three feet high whose tops are covered with an iron plate. The plate is large, flat, square or round, and so positioned that the side located over the mouth of the oven is slightly inclined upwards so that the roaster who stands in front of the opposite side of the plate is protected from the fire and can properly turn the roasting leaves, there being no opening around the edge of the plate through which smoke may escape

2. a low and very long table (several in a larger roasting-house) or

rather a structure of stone or wood in the shape of a table covered with woven mats of fine reed on which the leaves are rolled

3. the workers themselves, both those who stand at the ovens and manage the roasting as well as those who sit cross-legged at the table and perform the rolling.

The leaves must be roasted when they are very fresh. If kept until the following day, they blacken while being roasted and lose most of their excellence. Consequently, leaves are brought to the roasting-house on the very day or on the evening of the day they are picked for immediate preparation. Gathered leaves must not be allowed to lie in a heap for a long period of time lest they acquire heat from fermentation and lose their powers. If this happens, the leaves should be spread out and continuously fanned until cooled.

The preparation itself is accomplished as follows. The roaster puts several pounds of leaves into the pan described above. The pan should be so heated by the fire underneath that the leaves placed in it, wet as they are with their own juice, crackle upon contact with the pan. The roaster zealously turns the leaves this way and that way with both hands lest they remain motionless in any one place and roast unevenly. Note the following: it is customary, particularly in China, that leaves from the first gathering, before they are roasted, be immersed in boiling water for one half-minute or for as long as it takes a person to count to thirty. This effects a double drain of the narcotic power that abounds in the young and juicier leaves. The fire in the oven must be regulated to a temperature that the handlers are just able to endure. The leaves must be stirred continuously until their heat, as it increases, nearly exceeds the toleration point for handling. At precisely this point, the leaves are removed with a wooden spatula shaped like an open fan. They are spread on a mat and left to the rollers. Since the leaves must be rolled on the spot while they are hot, each roller at once takes the modest portion in front of him. The leaves are rolled with both—as I said earlier—palms, not haphazardly but uniformly, to effect precise and identical curling. The pressure of the rolling process extracts a yellowish-green juice that makes the hands almost unbearably hot. Despite the discomfort, the rolling must be continuous until the leaves cool. For just as the curling will only succeed under proper conditions of heat, so the curls will last only if the leaves cool under the hands. The quicker the cooling, however, the better, since the leaves retain tighter curls. Hence cooling must be promoted by continuous fanning. Upon cooling, the leaves are returned to the roaster—according to the instructions of the project director, who meanwhile was occupied with roasting other leaves to be curled—for a second roasting until all their moisture is removed. The roaster on this occasion stirs the leaves gently this way and that way, not vigorously as on the first roasting. His object is to prevent the curls from unwinding, but there is no technique capable of stopping many leaves from opening up again.

After the second roasting, the leaves are removed from the pan and given to the rollers for a second and equally careful rolling. Then, if the leaves are properly dried, they are stored; if not, they must again be subjected to roasting and rolling. During second or third roastings care is exercised to lessen the heat from the fire in proportion to the loss of moisture. If the fire is too hot, the leaves burn and blacken. Fastidious persons repeat the roasting and rolling five or even, if they have the leisure, seven times. Each time they turn and roast the tea for a shorter period and over a gentler flame, so that it dries gradually and in easy stages. This preserves the desired lively and even greenness that is lost by quicker, excessive, and uneven roasting. To preserve the greenness it is also necessary after each roasting to clean with boiling water the pan of the liquid residue that is drawn from the leaves and is inclined to stick to the sides and spoil the tea leaves added. When sufficiently roasted, the leaves are placed on a mat spread over the floor, where they may be classified as were the raw leaves before preparation. Now the collection of prepared leaves undergoes a sorting wherein the coarser, unevenly curled, and overly roasted leaves are assigned to a lower class. The leaves of Ficki tea should be roasted a little drier so as to be fit for grinding and reduction to a powder with less effort. These same leaves, if very tender and very young, are, after immersion in boiling water, normally roasted or at least dried on coarse paper spread over burning coals; due to their small size they are not rolled at all. Country people roast their leaves in earthen kettles a few times and with no precision. Consequently, their tea does not achieve the highest quality, and they sell it thus unskillfully handled to the common people at a much reduced price. When tea has been stored for some months, it must as a precaution be taken from its containers and again roasted over a very gentle flame. In this way if any moisture remained or was contracted during the rainy season, it is removed and the tea may again be stored without danger of spoiling.

The masters of this art have a complaint about their profession. Although nothing, they assert, is cheaper in this realm than tea, still no work is more fatiguing, and indeed it must be carried out contrary to nature's law—during the night when men should rest.

VIII
Preservation

As soon as roasted tea has cooled, it should be stored and carefully protected from the air. It is not surprising that prime importance in the process of preservation is attached to protection against the air that in a climate warmer than our own rather cold one, more easily dissipates the quite subtle excellence of tea. I do believe the tea brought to Europe has indeed lost the very subtle element of salt; for in Europe I do not find that grace of savor nor refreshing quality that tea drunk in its native land has furnished me. The Chinese use boxes made from sheets of cheaper tin. Larger boxes must be strengthened with an outer casing of firwood whose openings are filled with paper on both the interior and exterior. Tea so packaged is shipped abroad to various places. The Japanese keep their stock of common tea stored in large earthen jars with rather narrow mouths. They prefer to keep the superior tea required by the emperor and the highest officials in murrhine or porcelain jars, especially, if they can be had, the precious little vessels celebrated for their antiquity and called Maats ubo. These vessels are credited with not only preserving but also enhancing the excellence of tea so that the longer it has lain in them the more expensive and superior it is considered. Pulverized Ficki Tsjaa is also kept in these vessels for some months with no loss of its powers. Tea placed in Maats ubo vessels is said to recover fully whatever excellence it lost from exposure to air. Hence, the wealthy strive to purchase and possess at any price one or two that enjoy the place of honor among the utensils found at luxurious tea parties. The excellence of these pots merits a fuller account that no one has yet published and that I here add.

Maats ubo means a true pot, as if one were to say the most outstanding of all types of pots. There exist ancient vessels so named and made from the finest clay on the island of Mauri ga sima, that is, the Island of Mauri. Mauri, so the story is told, was completely submerged by the gods because of the immorality of its citizens. There is not a trace of the island left except for some rocks, which are visible when the sea subsides. Mauri was situated near Teyovaan or the island of Formosa. Its location is indicated on hydrographic charts by dots and asterisks noting a bottom infested with shoals and rocks.[14]

This is the account the Chinese give of the island's fate. The island of Mauri was celebrated among the ancients for its rich soil that among other things, yielded superb argil[15] for potters to fashion murrhine— today called porcelain—vessels. Profit and wealth from these brought luxury to the inhabitants. From luxury, vice and a contempt for religion arose, which so offended the gods that they by an irrevocable decree condemned the island to destruction by flood. A dream sent from heaven revealed the decree to the island's ruler, a pious man of upright life named Peiruun, with this admonition: to save his life he should embark on his ships and take flight as soon as he saw the faces of two idols placed before the temple entrance turn red. Gigantic statues of the two gods, called In-Jo, Ni-Wo or A-Wun, were made from wood. The one is believed to preside over generation, the other is believed to preside over destruction. The one denotes heaven and the active principle; the other, earth and the passive principle. The one opens and gives; the other closes and receives. Each image has a lion's face with a crown on the forehead and holds a short staff of imperial authority entwined by a snake. The scepter of In is held erect in his right hand; that of Jo is held in his left hand and pressed against his thigh. The buttocks are covered with a cloth, and as the wind blows, the drapery moves about the body. Chest and limbs are bare. The one's mouth is wide open; the other's is closed. They receive their names from their functions and gestures. The one is called In, Ni, and

A by the learned and Rikkisi woo by the common people. The other is called Jo, Wo, and Wun by the learned and Kongo woo by the common people. These, then, were the idols placed (as is still the custom) as ornaments in the entrance of the great temple and destined to reveal the coming destruction.

The king publicized the impending punishment and the sign of the disaster but created among his subjects nothing but derision and contempt for himself as a man of superstition. Not long thereafter some rascal bent on mockery approached the idols by night and, unseen by anyone, painted both faces red. When advised of the change of color, the king did not credit the occurrence to a wicked human but took it the fated miracle. He anxiously fled from the island with his family and hasted with oars and sails to Foktsju, the nearest province of southern China.[16] After the king's departure, the mocker and his accomplices feared nothing sinister from their own act of insolence. But they as well as the whole throng of unbelievers were swallowed up by the sea as the island submerged with its potters and splendid array of murrhine vessels.

The event is celebrated with an annual festival by the southern Chinese. On this day they row their boats swiftly over the waves and bays of the sea as if they were fleeing and call out Peiruun, the name of the ruler, as one who has been lost. The same festival has been introduced into western Japan. When the sea subsides, the vessels are searched for at the bottom of the water. They must be separated carefully from rocks to which they are stuck in order to avoid breaking. The vessels are disfigured all about by attached shells that the polishers scrape off, leaving only a small portion to guarantee authenticity. The vessels are transparent, very thin, and of a greenish-white color. Their usual shape is that of a small wine jar or jug with a short, narrow neck, almost as if the vessels had been originally made to preserve tea. They are imported and sold in Japan (quite rarely) by merchants from Foktsju province who acquire them from divers. The cheaper ones cost about twenty taels; the average ones cost one hundred or two hundred taels, the larger ones, if unblemished, cost three, four, or five thousand taels, and no one but the emperor would dare to purchase the ones so expensive and fine. The emperor is said to possess by way of inheritance from his ancestors and predecessors a quantity that no amount of gold could buy and that consequently enjoys the place of honor in the treasury. Rarely are the vessels free from breaks or cracks. Polishers are so skillful in repairing these that a blemish will be undetected by the sharpest eye or any technique unless the vessel is boiled in water for two or three days so as to dissolve the glue. So much for Maats ubo, the superb containers for tea leaves.

Ban tsjaa tea does not as easily lose its flavor from exposure to air, because it is stronger than the other teas, although it is also many grades inferior. Therefore, it scarcely requires the same meticulous preservation. The country people keep it, as they do all other tea, in straw baskets shaped like casks. They keep these just below the roofs of their homes so that the smoke may gently encircle them, for smoke is believed to fix and preserve the goodness. They claim that tea is rendered more savory when stored with flowers of the *Artemisia vulgaris*[17] or young leaves of the Sasanqua plant.[18] Other odoriferous substances are not equally suitable for tea.

IX
Use

Tea is prepared for drinking in two ways. The first, the Chinese way, is infusion. Over whole leaves one pours boiling water that is then sipped as soon as it has absorbed the virtue of the leaves. This method, as it was brought to Europe from China, is known even to the most ignorant men. It does not require an explanation on my part. The second method is grinding. On the day before use or on the day itself, leaves are ground into a very fine powder with a hand mill made of a black stone that they call serpent stone.[19] The powder is mixed with boiling water to the consistency of a rather thin pulp and consumed by sipping in the same way as the first type of prepared tea. This is called Koi tsjaa, that is, thick tea as opposed to thinner tea produced by infusion. The use of this tea by the wealthy, the important, and the noble is standard in almost the whole of Japan and unique to this country. It is prepared and served as follows. The powder, which is enclosed in a box together with an array of vessels, is brought into the room where tea drinkers are gathered. Next, the box is opened, and a portion (about the amount you could take with the point of a larger knife) is placed with a tiny and highly polished spoon into each dish filled with boiling water. Then it is stirred in a circular direction and thoroughly mixed with a curiously dentate utensil until it foams. At this point it is served hot to the guests for sipping.

There is also a third method: boiling. This is practiced by the common people in the cities and the country who keep tea in readiness all day long in the following way. Upon arising before dawn, one member of the household hangs an iron kettle filled with water over the fire. He places, according to the number of people in the house, two or three or more handfuls of Ban tsjaa leaves in the water either before or after boiling, according to one's pleasure. Then he inserts a basket of a shape and size that fits snugly into the hollow of the kettle. The basket presses the leaves to the bottom and prevents them from hindering the removal of the beverage. This kettle supplies the entire household with a beverage to quench its thirst. Everyone may hurry to the kettle when he has a mind to. In front hangs a dipper. Cold water is also kept close by so that if one does not have the time to sip leisurely, the tea can be diluted and cooled to quench thirst with large draughts and little waste of time. Others dispense with the basket and instead boil tea in a bag with the same results. Ban tsjaa tea alone is suitable for boiling, since its virtue is quite enveloped by resinous particles and cannot be drawn out without sufficient boiling. Once, however, the water receives this virtue, it does not easily lose it.

The knowledge and practice of both preparing tea and serving it to a group of guests are quite special. More importance is attached to style than to technique of boiling and preparing. They call this art Sado and Tsjanoi.[20] Masters of this art teach children of both sexes what they term Tsjanosi, that is, to belong to a symposium of tea lovers and to serve tea in a graceful and praiseworthy manner. It is the same with Europeans when we learn the fine points of carving and apportioning food, of sports, and of similar practices. Impoverished workers, especially in the province of Nara,[21] sometimes cook their rice (the mainstay of the nation) in tea prepared by infusion or boiling. This rice, they believe, is so much more nutritious and satisfying that one portion equals three portions of ordinary rice. The cheapest variety of tea also has an external use. When it has lost its strength over a period of several years, it is used for dyeing silks to which it lends a chestnut color. For this purpose a large amount of tea is shipped from China to Gusarrata almost every year.[22]

X
Virtues and Vices

I have noted that tea leaves have the power to inebriate or to disturb the animal spirits. This force should be driven out by gentle and repeated roasting, although the process cannot completely remove a certain element that is hostile to the brain. This element, however, weakens spontaneously after ten or more months. After this amount of time, rather than disturbing the animal spirits, tea pleasantly stimulates them and soothes their organs so as to produce the effect of an exhilarant. Tea less than one year old has a very delicate taste, but if taken in large quantities, it causes the limbs to tremble. Tea is at its finest, most delicate, and most exhilarating after one year. No Japanese drinks tea less aged unless it is mixed with an equal amount of older tea.

We can sum up the virtues of tea in a few words: tea clears visceral obstructions, purifies the blood, and in particular washes away the tartarous matter of calculous concretions. So effective in this respect is tea that I have never encountered a tea drinker in Japan who suffered from stones or gout. Europeans could hope for the same benefits from drinking tea were it not that they inherit an element that gives rise to these diseases and a stubborn inclination to them that are nourished by tart wine, beer, and lixivial meats.[23] Too little value is assigned to tea by lovers of the native beer brewed from rice and called Sampsu in Chinese and Sakki in Japanese. They deny that tea has a power any more special than that of correcting the harshness of water and uniting a group of friends for conversation. Such people, however, frequently suffer from traveling arthritis and strangury.[24] Those who substitute *Veronica*[25] and *Myrtus brabantica*[26] for tea are greatly mistaken. To this day there is no other plant known to mankind whose infusion or decoction, when taken in such quantities, is less disturbing to the stomach, is passed with equal ease, and is so quickly able to stimulate and refresh the weary spirits. One might find more profit in seeking similar benefits from plants that are rejected by us for their malignant properties but really require only the right preparation by correction and tempering. Europeans, however, are quite ignorant of the method and indeed are hostile to any such approach to forbidden plants. The mere mention of it is enough to invite the stigma of witchcraft on the honor of one who would devote himself to such abstruse powers. The ingenious Brahmins know the method. Long experience has taught them how to prepare and correct the Datura, the poppy (whose superb juice eminent jurists include among poisons; Gothofr. ad L. 3ff. ad *L. Corn. de Sic*)[27] and similar local plants possessing narcotic properties. The Brahmins produce agents for inducing forgetfulness, for dispelling sorrow, for exhilarating the spirits, and for stimulating all sorts of ideas in men. These are usually administered in the form of an electuary.[28]

The Japanese list the following vices of tea. All tea drinking suppresses the efficacy of medicines. In cases of (endemic) colic extreme care must be taken to avoid its damaging effects. Just as tea brewed from fresh leaves generally disturbs and injures the entire fabric of the brain, so according to precise experiments is it extremely harmful in cases of inflammation of the eyes. While questioning Chinese physicians, I learned the following from an elderly one. A person who drinks strong tea all day long will destroy the radical principle of life that consists in the balance of hot and cold (or dry and moist). The same results are obtained but by the opposite process if one daily gluts himself at every meal with fatty foods and especially pork, a most common dish among the Chinese. If one joins these contrary elements, the effect, far from injurious, is exceedingly beneficial to health and longevity. This, they said, became clear to them in the case of a certain woman. Weary of her emaciated and impotent husband, she wheedled from her physician a means of doing away with him. The physician ordered the woman to stuff her husband with pork and all types of fatty foods in order to extinguish within a year the meager light of life remaining in that shadow of a man. The treacherous woman, not content with one means, consulted another physi-

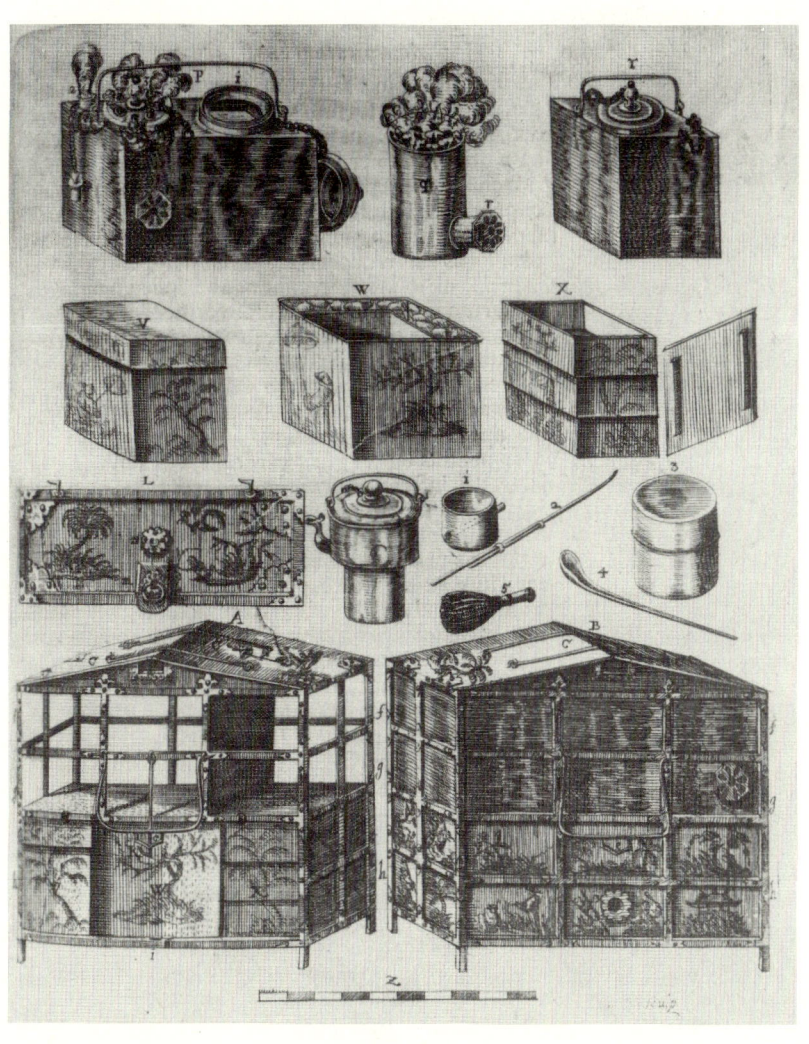

Plate 10. *Tea Apparatus*

cian. He advised her to serve her skeleton of a husband large quantities of strong tea and thereby achieve her design within a year. In order to free herself the more surely and quickly, she employed both means. The combination, far from speeding the condemned man on the way to his coffin, caused him to regain his strength, fatten his constitution, and completely recover his health.

As I write, a similar case involving a faithless woman comes to mind from Ausonius. Bent on killing her husband, she gave him a poison; to double the lethal power she also gave him quicksilver, which saved his life. An elegant epigram (pardon the digression) by the poet Ausonius Gallus describes the events:

A faithless wife gave poison to her jealous husband,
But was not too sure that it would kill him.
In she mixed a deadly dose of quicksilver,
The doubled force to speed sure death.
If one takes them separately, by themselves they are poisons;
If one drinks them together, he imbibes the antidote.
Hence, while noxious draughts battle each other,
Lethal force yields to healing force.
Directly, they seek the stomach's empty recesses
By the easy path known to swallowed food.
How great the loving care of the gods!
The too cruel wife is an asset,
And, when the fates wish, two poisons an advantage.[29]

XI
Instruments and Vessels Used for Making and Serving Tea

In order to complete this account of tea, I am pleased to furnish illustrations of a portable and compact container together with all vessels and instruments necessary for making and drinking tea. This is the type of container carried by the Japanese for their refreshment when traveling. Indeed, they never journey without this touch of sociability.

A and B: Two views (A front, B rear) of the portable machine complete with brass hooks, hinges, braces, and nails. Except for the folding doors (at the top), it is all wooden and varnished.

cc: Two brass folding doors meeting at the very top [of A and B] and secured by two long braces.

d: An opening just beneath the very top of the machine aligned with the opening on the ends (ee) [low center] of the handles when raised. A pole inserted through all these openings permits the machine to be carried on the shoulders.

fg., fg: The upper level of the machine. It contains two brass vessels (P and T) plated with tin on the inside. The vessels are used for storing and boiling water. They are removed by opening the folding doors (cc) that form the machine's top.

gh., gh: The lower level of the machine. It contains three rows of wooden boxes (V, W, and X), which are elegantly varnished on the inside and outside. The boxes contain items necessary for drinking tea.

i: The opening, into which the bolt of the hanging door (L) is inserted.

K: A long brass hook that is fitted to an eye to hold open the top-hinged door.

L: The wooden top-hinged door, removed from its hinges. It closes the lower level on the side shown in A to prevent the boxes from falling out. On the door can be seen: the lock or bolt (m) that fits into the opening (i) noted above [bottom center]; and the little eye (n) [center] that receives the above noted hook (K) in order to hold the door open while the boxes are being removed.

o: An opening in the rear of the machine (shown in B [bottom center]). To facilitate removal of the boxes, a finger is inserted into the hole and pushes them out.

P: The larger brass vessel in which water is boiled. On top are three openings, each with a cover. Cold water is put in the first opening (1); hot water is poured out from the second and small opening (2); coals are placed in the third opening (3), which leads to a wind oven inside. I show one cover hanging down with its underside visible so that one can see the two edges designed for a tighter fit. The wind oven (q) is cylindrical and made of brass. It is placed in the water to heat it and is attached to the larger vessel that houses it at the opening on the top and the mouth on the side. (r) is the mouth of the oven. Through it passes air to kindle the coals. (s.s.s.) are the breathing holes on the cover that allow smoke and soot to escape.

T: The smaller brass vessel or container for cold water. It has a cover like the one I described above.

V: A wooden box containing the utensils and small vessels for making and drinking tea. Each item is numbered I [1], 2, 3, 4, 5, or 6.

1. A dipper fitted with a tube running through it.

2. The stem or handle of the above. It is removed after use.

3. A small container filled with whole or ground tea.

4. A spatula or spoon for taking out ground tea.

5. A mixer with which Ficki tsjaa is mixed before being sipped.

6. A brass vessel used for pouring out tea. Its lower part, which is plated with tin, is placed into the opening of vessel P, where the steam or hot water keeps the tea from cooling. It has a tight-fitting cover of the type described above.

W: A second and larger wooden box with two compartments. The outer compartment contains the coal and materials for making a fire. The inner compartment contains various dishes for drinking tea and whatever else one may wish.

X: Three smaller boxes that are alike. They rest one on top the other with the cover of the topmost box opened to reveal its underside. Here are kept various edibles usually served to guests along with tea.

Z: The scale for the illustrations. One can obtain the size and proportions of the individual parts and vessels should he wish to construct a machine like this one. The scale contains one saki or ten sun, which are equal to about one of our feet.

Notes

1. Wilhem ten Rhyne (1649–1700), Dutch physician and botanist, served on Deshima from 1674 to 1676. In 1678 at Danzig his pioneering study of tea was published in Jacob Breyn's *Exoticarum plantarum centuria prima.* See also Observation IV, note 16.

2. *Cerasus hortensis fructu acido*: cultivated sour cherry tree.

3. *Euonymus vulgaris granis rubris*: the *Euonymus* is a genus of shrubs or trees of the order Celastraceae, which are sometimes spinous or climbing and have seeds with scarlet arils.

4. Darma (Daruma), also Bodhidharma, called the White Buddha by the Chinese, came to Canton in A.D. 520 from India with the sacred bowl of the Buddhist Patriarchate. He believed that real merit was found in a combination of purity and wisdom, and he taught that religion was not to be learned from books but that people should seek and find the Buddha in their own hearts. Darma crossed the swollen waters of the Yangtze on a reed, a feat that inspired Chinese painters and poets.

Although Chinese legend relates that tea (*Camellia sinensis* or *Thea sinensis*) was discovered by the Emperor Shen Nung in the third millenium B.C., the first reliable mention occurred in the fourth century A.D. The use of tea spread in China and then to Japan with the help of Buddhist priests campaigning against intemperance.

5. satori: in Zen Buddhism, a state in which the follower achieves spiritual enlightenment.

6. sphacelus: process of becoming gangrenous or decayed; the decay itself.

7. palm: measure of length; either the breadth of the palm of the hand (three or four inches) or the length of the hand (seven to nine inches).

8. Tea is not cultivated in Europe. Turkey is the only Mediterranean country in which tea is grown. Tea came to the attention of European travelers during the sixteenth century. The Venetian Giovanni Battista Ramusio first mentioned tea as *chai catai*, or "tea of China," in his *Voyages and Travels* (1559). By the early seventeenth century, tea was imported to Europe by the Dutch East India Company. In 1658, the *Mercurius Politicus* published the first newspaper advertisement: "That Excellent and by all Physicians approved China drink, called by the Chineans Tcha, by other nations, Tay, alias Tee, is sold at the Sultaness Head Cophee house in Sweetings Rents, by the Royal Exchange, London."

9. catty: a measure of weight equal to one and one-third pounds.

10. Udsi: a town on Lake Biwa, near Miaco (now Kyoto).

11. A sja mome is a silver Maas, a unit of money. A tael equals an ounce of silver, so its value depended on the value of the metal at any given time. A mace is 1/10 of a tael. stiver is a coin equal to 1/20 of a guilder.

12. Mt. Hiei, once a center of Buddhist power, northeast of Kyoto.

13. murrhine: porcelain earthenware.

14. Mauri is possibly one of the Ryukyus obliterated by a volcano. These islands stretch from seventy-three miles northeast of Taiwan to a point eight miles south of Japan and include Okinawa.

15. argil: clay.

16. Foktsju: Fukien in China, just across the Straits of Formosa.

17. *Artemisia vulgaris*: a large genus of plants including mugwort and used for moxibustion (see Observation XII).

18. Sasanqua: a shrub (*Camellia sasanqua*) with fragrant evergreen leaves and white or red flowers.

19. serpent stone: Kaempfer discusses serpent stone or snake stone in Observation X, where he refutes the belief that it was an antidote to the venom from snake bites and rejects as well the assertion that snake stones are obtained from the heads of snakes.

20. Sado and Tsjanoi: sado means tea ethics;" tsjanoi or cha-no-yu means "tea of hot water." The tea ceremony, a traditional and highly stylized ritual, traces its origin to the principles of Zen with emphasis on the beautiful in daily life.

21. Nara: the city is located close to Osaka.

22. Gusarrata: Gujarat on the Bay of Cambay, India.

23. lixivial: containing alkaline salts.

24. strangury: painful and interrupted urination.

25. *Veronica*: a genus of plants and shrubs with pink, white, blue or purple flowers in spikes.

26. *Myrtus brabantica*: a myrtle shrub.

27. Gothofr. ad L. 3ff. ad *L. Corn. de Sic*: Jacques Godefroy (1587–1665), the most celebrated jurist of his time, is best known for his *Codex Theodosianus* (4 vols. [Lyon, 1665]), to which he devoted thirty years of labor. Kaempfer's

reference is to the *Lex Cornelia de sicariis* (81 B.C.), which includes poisoners among those guilty of homicide. Godefroy, in commenting on the Cornelian law, lists poppy juice as a poison.

28. electuary: a medicine made into a paste with honey or syrup.

29. Ausonius: Decimus Magnus Ausonius (c. 310–395), the Latin poet and professor of rhetoric at Bordeaux. Ausonius tutored the emperor Gratian, who made him consul in 379. Ausonius converted to Christianity. The translation of epigram 3 of *Epigrammata Ausonii de diversis rebus* is from Hugh G. E. White, *Ausonius*, Loeb Classical Library (London: Heinemann, 1919–21) 2: 157.

Observation XIV
A Defense of Ambergris

I
The Substance of Ambergris

Since my lengthy discussion of the previous substance went beyond the limits of mere observation, by way of compensation I will now offer a shorter presentation of a precious and most desirable bitumen. What is sweeter to smell than ambergris?[1] From the hidden depths of its bosom the ocean with sparing hand bestows it upon the human race. For it is nature's way to conceal energetically and to impart parsimoniously those substances that are especially noble. Authorities differ in no small ways on the questions of the origin and the nature of ambergris. Some believe it to be a type of bitumen; others, a product of the earth; others, a sea sponge; others, whale excrement; others, bird dung; and there are additional identifications. Of all the theories known to me, none is more ill-founded than the recent one based solely on similarity of substance or smell. Jean-Baptiste Denis (*Journal des scavans de l'An 1672, Conference Seconde*)[2] concludes, "Ambergris is a compound of honey and wax, deposited on the seacoasts by bees. It is vaporized by the heat of the sun and drops into the sea. Then it is processed by the action of the waves and salt, and transformed into this noble substance." This conjecture is foolish and groundless! It aims to please solely by its novelty, and with the support of the great Prince it seeks to dismiss the opinion approved by those who have examined the matter in depth. These experts have already demonstrated that ambergris is a type of bitumen produced by the earth, or what amounts to the same thing, a subterranean fat that is boiled into a bitumen and deposited through underground veins and channels in the depths of the sea, where it is condensed by the sun and salt.

I will now proceed to free ambergris from this new and erroneous theory by some arguments that I derived from collectors of this coagulum, from curious observations by the Chinese, from Japanese whalers, and finally from personal inspection of the regions and shores that produce ambergris.

1. Ambergris is collected in several places where there are no bees either around the shore or in the interior of the region. On the contrary, many regions abound in bees, while no ambergris is found on their shores.

2. Fishermen who gather the edible nests of the halcyons from the coastal rocks around China and Java (these are nests that sea swallows build with pieces of holothurian),[3] say that they have never seen beehives hidden among coastal rocks, the very hives Denis imagines are torn loose by the waves. Indeed, in its providence nature has taught bees to avoid the shores of the sea and stormy places.

3. Honey, wax, and honeycomb, when mixed with a liquid, do not unite into one substance, but separate and are dissolved.

4. Wherever in the world honeycombs together with their honey are thickened by fire, they produce a coagulum of an identical substance. The many varieties of ambergris, each proper to its own region, are on the other hand due to the differences of their subterranean veins. Consequently, by examining ambergris, experts can identify the coast on which it was found, just as connoisseurs of wine recognize the region of the wine by tasting it. Some varieties of ambergris resemble rather crude bitumen, asphalt or dry black naphtha, and to varying degrees are white, light, and clear. Still other varieties are often so spongy that the learned Scaliger followed Serapion in taking, quite reasonably, ambergris for a marine fungus.[4]

5. Ambergris, when freshly deposited from the depths of the ocean, is soft and has the appearance of cow dung. At the same time, it has an odor of burning that is entirely foreign to the production of honey.

6. Quite frequently we find shiny black shells and other submarine fragments in ambergris. Sometimes we find particles that ambergris, when rather soft and freshly deposited on shore, can envelop. But we never find bees, honey, or honeycomb. The distinguished Denis was deceived when he believed reports that honeycombs with wax and honey (why not with bees too?) are found in ambergris. Recent French writers are likewise deceived when they believe Denis.

7. Occasionally, large lumps of ambergris far surpassing the capacity of a beehive are found. I pass over the pieces of monstrous size which Garcia da Orta cited.[5] For me, lesser masses witnessed by myself are sufficient. During my sojourn in the Orient a mass of grayish ambergris was discovered on the coast of the province of Kinokumi in Japan. It weighed more than one hundred catties,[6] or about one hundred and thirty Dutch pounds. When no single buyer could be found for the whole mass, it was divided and cut into four pieces in the shape of a cross. Three pieces were sold. The remaining fourth part was offered for sale to me upon my arrival. From this one piece it was easy to reconstruct the great size of the whole. A mass larger than this one was sold in 1693 by the King of Tidore[7] to the Dutch East India Company for 11,000 Joachimsthalers.[8] The following year it was dispatched to the Rarity Room at Amsterdam. Its weight (carefully measured) was 185 Dutch pounds. Moderately gray and of high quality, it had the shape of a tortoise lacking a head and a tail. It was sold with the agreement that the price be repaid should adulteration be discovered. The distinguished Professor Valentini of Giessen has an illustration of it in his *Museum museorum*, book 3, chapter 28.[9]

II
The Attributes of Ambergris

The following remarks comprise what I have learned about both the adulteration of ambergris and the signs of its goodness and virtues.

When freshly deposited on shore, ambergris resembles a lump of meal and is most susceptible to adulteration. As those engaged in adulteration themselves confessed to me, nothing blends better with ambergris than pulverized rice husks, which lend lightness and a grayish color. But the buyer recognizes such fraud with the appearance of worms and decay. The adulteration of ambergris with storax,[10] benzoin,[11] and other fragrant substances is somewhat more difficult to detect. Pseudo-ambergris, a composite of tar, wax, rosin, storax, and similar substances, is quite easy to detect by touch, appearance, or odor alone. Both varieties were frequently offered for sale to me.[12] Those who discover pieces of fresh ambergris squeeze them into one larger mass. If a mass is too large and ill-shaped, they compress it into a more suitable form. Consequently, ambergris often comes in the shape of a ball, which increases density without detracting from goodness. The usual and most reliable indications of quality are obtained by placing a few grains on a red-hot plate. The smoke reveals adulteration; the paucity of ashes, genuineness. The Orientals beyond the Ganges employ a kobang, a thin gold piece of money of oblong and rounded shape, kept in readiness for this purpose.[13] They put a scraping of ambergris on the piece of money, which is then placed over hot coals. The Chinese consider the best variety of high quality ambergris to be the one whose scrapings, when immersed in hot water and covered, more evenly dissolve and liquefy.[14] For this purpose, I have seen them use murrhine dishes, from which they sip tea, and the ornate ware traditional at luxurious banquets. The ambergris found in the intestines of whales is considered the most inferior type, because its powers have been exceedingly diminished by its location. A type of whale, called Mokos, which is three or at most four fathoms long, is caught in the waters around Japan. Ambergris is very often found in its intestines. If a grumous substance resembling lime appears while the intestines are being removed, this is a sign that ambergris is concealed within. The type of ambergris that is found in the innards of whales as well as the type of ambergris that live whales discharge into the sea along with excrement are so common that the Japanese regularly call ambergris Kusura no fuu, or whale dung.[15]

The seas around the southern shores of the East Indies frequently deposit pieces of strange fat that deceive their finders by an external resemblance to ambergris. A lump of this kind, cast up on the shore of Luzon, was pressed upon me as very fresh, grayish ambergris. It was whitish, fungous, brittle, reeking, and oily and smelled like rancid lard. Having judged it to be whale fat altered by a prolonged exposure to salt, I rejected it. I have an ill-shaped lump from the shore of Banda[16] that weighs about three pounds. It was a gift and supposedly real ambergris. But I take it to be a tallow of the kind to which Schröder gives the name *Ambra subalbida* (whitish ambergris) and which is commonly called spermaceti.[17] While floating on the sea, it collected, by the action of the waves, somewhere (perhaps on a rock) and was melted by the sun into one lump.

I have found three types or collections of this so-called spermaceti. The first type, which floats in northern waters and is gathered with wicker baskets, has long been known from the reports of eyewitnesses. Large quantities of the second type are extracted from the head of a certain whale, called *orca*[18] in Latin and potvis[19] in Dutch, according to Bartholin,[20] Worm,[21] and those who sail to Greenland to catch these whales. The third type is found around the Ferre or Fero Islands situated to the north of Scotland.[22] The local fishermen, the poorest of men, collect it from the outside of a fish that has a long head and that in their language is called Buskoppe. Because of the abundance of the fish, the northernmost of the many islands has also been called Buskoppe. My information on this last type is not derived from any book but from the account of a trustworthy man who suffered shipwreck on those islands. He swore to me that he had both seen the operation and had himself engaged in it during the six months of his stay. He said that the fish was larger than a man and that its huge head, especially around the jaws, was covered with a large amount of this mucous and fatty substance, which the fishermen scrape off, cleanse, and protect from growing rancid by strong lye and exposure to the sun. Fishermen of the Indian seas remove from the head of the shark (the most deadly sea-bandit except for the crocodile) the very white substance that they sell as spermaceti. This substance is not to be confused with the brain, which I found to be tiny and located in the center of the head. This substance excellently furnishes the diuretic power of spermaceti, but as mere powder, and lacking mucilaginous fat it must be excluded from the list of spermaceti. The coagulum from Banda that I possess has all the characteristics and qualities of purified spermaceti. It appears to belong to the first type discussed above, provided it is reduced again to the bran-like substance that spermaceti resembles. I have on occasion substituted it for spermaceti with success.

III
Yellow Amber, or Succinum

Natural scientists, with understandable error, consider amber a type of ambergris. Amber, like ambergris, is a subterranean fat hardened by salt and air, but it is translucent. While I have seen amber collect on the Prussian coast by the movement of the waves, I have also seen it mined in Prussia from the veins and bowels of the earth. But the mined variety, which is often soft and brittle, is placed in seawater to harden. Had Denis known this, he would not have transported amber by so arduous a route from the forests of Sweden to these shores. The nations of the Far East set a much greater value on amber than on ambergris. Indeed, they far prefer it to our precious stones (except for red corals), for which they have no use or esteem. They value as much finer and dearer the clear yellow amber. While Europeans have very little esteem for it, the Easterners attribute to it antiquity and the highest perfection. For them, other varieties of amber have little appeal. I attempted to persuade them otherwise, but they merely laughed at me. I achieved as much as a person would who endeavored to convince us that silver is more precious than gold. The dark people of the East have no use for the ambergris found there. Europeans have greater use for it in medicines. But the Mogul kingdom, Persia, and Arabia make the greatest use of it for confections and all sorts of delicacies. The Chinese, Japanese, and Tonkinese collect ambergris almost for the sole purpose of mixing it with odoriferous compositions in order to fix and heighten their sweet smell. Without the addition of ambergris, the sweet smell tends to weaken and disappear. Ambergris, which has little odor of its own, very effectively prevents this. There is no point in my discussing the virtues of ambergris, for they are quite well known. I would like to note one secret power of ambergris, which I learned from a very knowledgeable Japanese physician.

Recipe: Wrap any quantity of crude opium in a linen cloth and suspend it in the steam of boiling water. The exudation, which collects about the small knot after one or two hours, furnishes the finest opium. Mix this with twice the amount of ambergris and make tiny pills. A small dose taken before bed will assure sexual potency.[23]

Notes

1. Ambergris is a concretion formed in the intestinal tract of the sperm whale. Possibly it results from an intestinal irritation caused by horny portions of the squid or cuttlefish on which the whale feeds. Ambergris is found floating on the sea or may be washed ashore. The word *ambergris* comes from the French *ambre gris*, that is, gray amber.

2. *Journal des scavans de l'An*, termed the "father of modern periodical literature," was founded in Paris by Denis de Sallo in 1665. For Kaempfer's reference, see Jean-Baptiste Denis, *Supplement du Journal des scavans des annees 1672, 1673, et 1674* (Paris, 1729).

3. Holothuroidea: a class of echinoderms to which the sea cucumber belongs.

4. Julius Caesar Scaliger (1484–1558). Scaliger notes three reported types of ambergris: (1) a bitumen pouring forth from the springs of the sea; (2) a species of marine fungus; and (3) a product of the whale (*Exotericarum exercitationum liber XV*, 104. 10, 161–63). Scaliger concludes that almost all the researchers who have examined this matter more closely think that ambergris comes from whales. Serapion (Yuhanna Ibn Serapion) says that ambergris fortifies and strengthens the brain (*Liber Serapionis aggregatus in medicinis simplicis, Translatio Symonis Ianuensis interpretate Abraam iudeo tortuosiensi de arabico in latinum* [Venice, 1497], 58).

5. Garcia da Orta, *Coloquios* (Goa, 1563), 1.1.

6. A catty equals one and one-third pounds.

7. Tidore (Tidor) is a mountainous island of the Moluccas. Nearly circular, it comprises an area of about thirty square miles. The Dutch conquered the island in 1654.

8. Prototype of the large, heavy silver coins of the sixteenth century, the Joachimsthaler was first coined in 1519 by the owner of the silver mines at St. Joachimsthal in Bohemia. Our word *dollar* is a modified form of *thaler*, a shortened form of *Joachimsthaler*.

9. Michael Bernhard Valentini (1657–1729), *Museum museorum* (Frankfurt, 1704).

10. storax: a resin derived from various styracaceous trees, especially the *Styrax officinalis*.

11. benzoin: a dry, brittle, resinous substance with a fragrant odor and a

slightly aromatic taste. It is obtained from the *Styrax benzoin*, a tree that grows in Java, Sumatra, and elsewhere.

12. C. Deite, *A Practical Treatise on the Manufacture of Perfumery* (Philadelphia, 1892), comments, " On account of its high price, ambergris is frequently adulterated, the commercial article being nothing but an artificial mixture of benzoin, olibanum, wax, and flour, with other substances, perfumed with musk . . . . In perfumery, ambergris is not so much used on account of its agreeable odor, but rather to make the perfumes more constant" (187–88).

13. This Japanese gold coin was first issued in 1609 and remained in general circulation up to 1870.

14. Modern tests for authenticity include: (1) dissolving a small amount of the substance in hot wood alcohol; as the alcohol cools, true ambergris will crystallize; and (2) pressing a wire or needle (previously heated in a gas or candle flame) into the substance, which, if true ambergris, will react in a special way.

15. This theory was resurrected in "An Account of Ambergris, by Dr. Schwediawer; presented by Sir Joseph Banks, P.R.S.," *Philosophical Transactions of the Royal Society of London* 73 (1783): 226–41.

KAEMPER, who has given us so many other faithful accounts in Natural History, seems to come nearer the truth with regard to the origin of ambergrise, when he says, that it is the dung of the whale; and that the Japanese, for this reason, call it, Kusura no fuu, i.e. Whale's Dung; but this relation, though founded on observation, has never obtained credit, and has been considered rather as a fabulous story, with which the Japanese imposed upon him, who had himself no direct observation to prove the fact. (234)

I think we may venture to conclude, that all ambergrise is generated in the bowels of the Physeter Macrocephalus, or spermaceti-whale, and there mixed with the beaks of the Sepia Octopodia, which is the principal food of that whale; and we may therefore define ambergrise to be the preternaturally hardened dung or faeces of the Physeter Macrocephalus, mixed with some indigestible relics of the food. (237–238)

C. P. Idyll comments:

This idea has been regarded by many recent writers with about the same amusement as the theories about Madagascar birds, yet Dr. Robert Clarke, the English whale expert, says, "I believe that Dr. Schwediawer was, quite simply, correct *** Local increase in water absorption by the large intestine and the chemical transformation enacted by the resident intestinal bacteria, probably each play their part in the formation of ambergris from feces impacted around some matrix of indigestible material." More observation is required on this point before any definitive statement about the exact manner of formation of ambergris can be made. The whale producing ambergris is not necessarily sick, as has been said, since perfectly normal whales have yielded lumps of this substance. Another theory, that only male whales produce ambergris, has also been shown to be false. ("Ambergris—Neptune's Treasure," *Annual Report of the Board of Regents of the Smithsonian Institution, 1959* [Washington, 1960], 377–83)

16. The Banda Islands comprise ten islands covering twenty square miles. The three largest islands (Great Banda, Banda Neira, and Gunong Alpi) are grouped around an inland sea that forms the Banda harbor.

17. Johann S. Schröder (1600–1664), *Pharmacopoeia medico-chymica, sive thesaurus pharmacologicus* (Ulm, 1641).

18. The word *orca* occurs in Pliny, *Historia Naturalis* 9. 5. 14. Pliny relates how Claudius (emperor A.D. 41–54) did battle with whales that entered the harbor of Ostia: "The emperor gave orders for a barrier of nets to be stretched between the mouths of the harbor, and setting out in person with the praetorian cohorts afforded a show to the Roman public. The soldiers hurled lances from the vessels against the creatures when they leaped alongside. We saw one of the boats sunk from being filled with water owing to a beast's snorting."

19. potvis: sperm whale.

20. Casper Bartholin (1585–1629) was professor of medicine at Copenhagen. While ill, he vowed to devote himself to religious studies should he recover. He did recover and kept his vow, becoming professor of divinity at Copenhagen. His son Thomas (1616–1680) was successively professor of mathematics (1647) and anatomy (1648). Another son, Erasmus (1625–1698), held chairs in mathematics and medicine at Copenhagen. In the third generation, Casper Thomeson (1665–1738), son of Thomas, taught anatomy at Copenhagen (*glandulae Bartholini* are named after him); his brother Thomas (1659–1690) was a scholar of antiquities.

21. Ole Worm (Olaus Wormius, 1588–1654), Danish anatomist, succeeded Casper Bartholin as professor of medicine at Copenhagen; see also Observation V, note 20.

22. The Faeroe Islands comprise a group of twenty-one islands in the North Atlantic between Great Britain and Iceland.

23. The medicinal value of ambergris is regarded as feeble. *The Dispensatory of the United States of America*, 24th ed. (Philadelphia, London, Montreal, 1947), notes: "Ambergris was formerly regarded as a cordial and antispasmodic, somewhat analogous to musk, useful in typhoid fevers, and various

nervous diseases. It has been used in doses of 0.3 to 4 Gm (approximately 5 to 60 grains)" (1318).

Observation XV
Kheif, or Keif:
Persian and Indian Intoxicants

I
Introduction

Eastern philosophers agree that nothing is more degrading and more contemptible than intoxication, the enemy of the human mind. Hence, cardinal doctrines of both pagan and Mohammedan religion solemnly warn against the use of wine and cider as liquors that profane the divine spirit in man. They are to be avoided more than snake or dog. You will, in fact, observe among the Turks, Persians, and Arabs of western Asia that individuals who have soaked themselves with liquor are punished with imprisonment and beaten on their bare feet until blood flows. The grape is not even grown between the Indus and the Ganges, nor is the method of eliciting spirits known. However, a species of wine, which the cut stem of the coco flower yields,[1] is collected at least for foreigners. Although the people on the far side of the Ganges are by no means abstemious, they still consider drunkenness an offense of the highest order. They have dreadful curses reserved for a host who would allow a guest at his home to get drunk and go forth into the streets to be stared at and mocked. However, despite the sanctity of the law or a reputation for baseness, the people of all these nations are drawn by a strange perversion of nature to the desire of drunkenness.

While they dare not use wine, they have learned to cheat the law with other ways and means of inducing insanity. A class of poisonous and deadly plants is found here and there in what we call the vegetable kingdom. We Europeans shudder at the mere names of these plants, and through timidity and scruples our physicians banish them far from catalogues of medicines. But the Orient with its longer experience has learned to prepare medicines from them. Small doses or somewhat larger ones after habituation, far from injuring a person's health with any malady, relieve the animal spirits from all darkness, the body from pain and the mind from depression; they very greatly exhilarate. They do not function in one and the same degree and manner because of the variety of species from which they are prepared as well as varying dosage and correction. Some gently soothe the animal spirits in the first stage of intoxication; others even disturb and stupefy them in a later stage; still others in the last stage excite and inflame these spirits to various types of insanity and sometimes even induce, by a strange transformation, the behavior of various beasts, depending upon the temperament of the subject and his mental inclination that emerges.

Here I will only review the three most impressive of the Persian intoxicants: nicotine,[2] cultivated white poppy,[3] and cannabis.[4] I will omit tea, coffee, and decoctions of the roots of the satyrion[5] and cinnamon along with rose water and sugar. I will also pass over milder exhilarants accepted as beverages at their parties. This power to intoxicate is called Keif, or Khief, as others write it. The term is famous and widespread in all India this side of the Ganges.

II
Persian Intoxicants

Nicotine

About a century and a half ago Portuguese merchants first introduced nicotine to all of the Old World and so to Persia as well. Its name internationally is Tobacco[6] and in the language of various nations, Tobak, Tabacoo, Tombak and Tembaku. The name is derived from the American island where an abundance of the herb was discovered.[7] Scarcely had the name of the plant been made known when at once its cultivation was eagerly undertaken everywhere, and the practice of smoking captivated the entire human race with astounding swiftness. If we deny that the plant is a species of *Hyoscyamus*,[8] this by no means removes it from the poisonous category, for vertigo, anxiety, and vomiting, which it induces in those unaccustomed to smoking, are reliable indications of its harmfulness. The experiments of Redi show that a drop of tobacco oil, when placed on a fresh wound, kills young fowl and in fact causes dangerous symp-

claudit orbiculus æneus, in fesquipalmarem diametrum
expansus, duos in medio permittens *tubulos* invicem ad-
solidatos, æneos; *Unum*, cujus inferior pars in ampullam
demissa, aquæ immergitur; superior recipit nicotianæ
cum impositis carbonibus retinaculum, infundibulo seu
buccinæ orificio simile: *Alterum* breviorem, cujus de-

missa extremitas aquam non attingit: superior incurva-
ta arundinem excipit longam, qua fumus attrahitur.
Tubulorum propago, proximè sub orbiculo, telâ xylinâ
arctè circumvoluta est, in eam crassitiem, quæ vitri orifi-
cium cum modicâ colli parte expleat atque claudat arctis-
sime: Ita evenit, ut ad suctum non possit nisi ex infundibulo fumus succedere; qui jucundo strepitu aquam pene-
trans, primò inanè vitri spatium occupat, inde per arun-
          Mm mm   dinem

Plate 11. *Persian Hookah*

toms in men.[9] I have seen porters around Kazan[10] in Tartaria empty with a few puffs a perforated ox horn filled with leaves and coals above them. The result was that, like epileptics, they dropped to the ground, overflowing with mucus and foam. Although the leaves are poisonous, nevertheless with experience men adapt to the smoke so that not only do its injurious qualities not harm them, but in fact, its quite beneficial salt removes serum from the recesses of the head and fills the brain with exhilaration.

To enhance the good results, the Persians draw smoke through a machine more than half-filled with water. The water absorbs the fetid sulphur harmful to the brain and transmits the smoke purified of its irritating malignity, cooled, and unspoiled. The machine, called a Khaliaan or Khaliuun, is a glass vessel one and one-half feet tall.[11] The opening of its long neck is covered by a small brass disc one and one-half palms in diameter that at its center allows for two attached brass tubes. The lower portion of the first tube descends into the vessel and is immersed in the water. The upper portion is attached to a container for nicotine with coals superimposed. The container resembles a funnel or the opening of a trumpet. The second tube is shorter, and its lower portion does not touch the water. Its curved upper portion receives a long stem through which the smoke is drawn. The portion of the tubes just below the disc is wound with cotton thread to a thickness that fills the opening of the bottle and a small part of the neck with a tight seal. Consequently, by suction nothing but smoke can exit the funnel. The smoke penetrates the water with an agreeable sound, fills the vacuum created in the bottle, and then by way of the stem reaches the mouth and the very lungs of the smoker. Indeed, the drawing is done not with the cheeks or the lips, as is common practice, but with the whole chest, which causes the smoke to spread through the lungs. If the herb is too bitter, it is cut up, immersed in water, and pressed to remove excessive roughness and bitterness. I have observed this practice among the Chinese and the Japanese. The Persian way of smoking with a machine was learned by the Hindustani Arabs, or the Great Mogul's Indians, and certain dark islanders who adopted the customs and religion of the Arabs. But for lack of glass, they use empty gourd shells instead of bottles.

The Turks, Chinese, and Japanese smoke in the European way with a tube inserted into a receptacle for burning tobacco. The dark heathen smoke without any device. The leaves are rolled in the shape of a cone whose base is lighted and whose apex is held between the lips and sucked. No people indulge in smoking more than the Persians, both men and women. Even while riding they scarcely give up smoking the Khaliaan. When the stem is a hindrance, they substitute a leather tube that is flexible, very long, and ingeniously protected with brass wires wound around it. A servant either on foot or on horse carries the machine itself along with a brazier. Since this Kheif is foreign rather than native and is common not only to Persia but the entire world, the brief notice of it here taken is sufficient.

White Poppy

A second Kheif for internal use is obtained from the poppy that Indians and Persians cultivate in gardens and fields in order to extract the milky juice by cutting the capsules. Europeans call this juice opium; Asians and Egyptians call if Afiuun and Ofiuun. The Persians out of respect call it, when prepared, Theriaki, that is, Theriac. For them this is the famous rest, cheer, and calm of the poets, that is, the medicine that gives the mind serenity, joy, and tranquility. For this reason we find that it has been given the three-word title, Andromache's Theriacal Antidote.[12]

In Persia it is harvested during the beginning of the summer by cutting crosswise through the exterior of nearly ripened heads. A knife fitted with five edges is employed. It makes five long cuts in one area. On the following day, juice emerging from the tiny cuts is cleaned away with a scraper and is collected in a small vessel attached to the abdomen of the collector. Then the other side of the head is cut in the same manner to elicit the liquid in the same way. But this process of collection must be undertaken a number of times in the same field because of the unequal growth and size of the heads. When a plant has an excess number of branches, superfluous heads are usually cut off beforehand; as a result the remaining heads grow larger and are filled with more potent juice. The tear of the first collection, called Gobaar, is the more excellent and has a greater ability to soothe the brain. Its color is white or pale yellow, which tends to darken from extended exposure to the sun and drying. A second harvest yields juice inferior to the former in potency and price, and usually is dark or blackish red. Some people undertake a third harvesting from which is obtained a very black tear of slight potency.

These are the basic steps in the preparation of opium. The opium is moistened with a very small amount of water and continuously and vigorously worked and reworked with a thick wooden spatula on a flat wood plate until the opium acquires the consistency, cohesiveness, and sheen of highly worked pitch. When the opium has been so treated long and carefully, it is at length worked considerably with the bare hand and at last rolled into short cylinders that are placed on sale. If a buyer wishes a portion, the cylinder is broken with tongs. Opium prepared by this process is called Theriaak malideh, that is, theriac prepared by milling, or even Theriaak afiuun, that is, opiate theriac, as distinguished from Andromache's theriac, which they call Theriaak Faruuk. This constant labor of preparation is done by small retailers, called Kheifruus (the German would be Trunken Krämere)[13] because they vigorously exercise their arms while sitting in marketplaces or at crossroads. This mass is often worked with sufficient amounts of honey instead of water to reduce not only the dryness but also the bitterness. Baehrs is the name reserved for this type. More outstanding is the preparation that entails, during stirring, the addition of nutmeg, cardamon, cinnamon, and mace, reduced to a very fine powder. Opium prepared in this manner is believed to offer superb benefits to the heart and brain. The species is called *polonia*, or, as others pronounce it, *folonia*, namely, Persian Philonium or Mesue. Others omit the aromatics and merely compound the mass with crocus and amber. Many people themselves complete the preparation at home for their own use in order to avoid being cheated by retailers with a small or large amount of admixtures.

In addition to opium prepared in three stages and swallowed only in the form of small pills, there is sold and also prepared in the household a liquid with the celebrated name Coconar. I think it is the *mêkonion*[14] of the Greeks and the Homeric *nepenthês*[15] that drinkers consume to their satisfaction at intervals of hours. Some people prepare the liquid with leaves boiled in plain water; others, with crushed heads soaked by infusion or placed over a filter with boiling water poured over them seven or eight times. Admixtures according to individual pleasure lend an agreeable taste.

I add a third type of opiate, a pleasureful electuary that intoxicates with joy. The same opium also constitutes the base of this electuary that is variously elaborated by dealers in unguents and physicians according to their individual ingenuity. Its purpose is to strengthen and exhilarate the spirits through varied ingredients, and consequently there exist varied descriptions of it. The most important and the most celebrated of these, whose discovery is credited to Hasjem Begi, is said to suffuse the mind of the taker with wondrous joys and to soothe the brain with fantastic ideas and delights.

Europeans assert that opium causes a horrible death if the dose exceeds one or a few grains; but the aforementioned peoples have from extensive use become so habituated that many suffer no injury from swallowing a drachma of it. Many evils arise from its abuse or prolonged use: the body becomes emaciated, strength drains away, the mind is saddened, the intellect dulled. Hence, you will observe devotees drowsy and almost incapable of speech—like blocks of wood—seated at opium parties. Addicts have often been brought to me to be cured of their bestial appetite for opium. Fees of one hundred gold pieces were promised, if I could protect them from injury and loss of life. There is no need for me to supply examples of the lust for opium; medical books are filled with them. Many people enjoy for dessert the more tender poppy heads preserved in vinegar; still others make various drinks from them, each according to his pleasure. I will proceed to Cannabis.

Cannabis

Persons who delight in a variety of intoxicants or are not quite captivated by the savor of opium seek the ecstacy of intoxication with the Cannabis plant. I will not here enter into the question of whether this is authentic native cannabis or a species of it called Bangue[16] (described in *Hort. Malab.* vol. 10, p. 119).[17] It appeared to me as like native Cannabis in all respects as one egg is to another—both the male

and the female. Further, I am prepared to believe that the excellence
it possesses is due to sky and soil. The Bengals of Gamron transported
seed from Isfahan in order to grow the plant. From the seed the
Cannabis plant did in fact grow with a sufficiently heavy odor but
minus the expected power.

No plant in Persia has the genuine excellence except that which is
cultivated in Komaereh or Luristan, in Kutelli Pyresen around Kazerun, and in the plain of the city of Isfahan.[18] It furnishes the joyous
parts: seed called Sjadoneh, pollen from flowerets called Tijeers, and
leaves called Baeng. The seed has a rather weak power; it is cooked
with foods and is an ingredient of emulsions designed to induce
intoxicating joy. Some men preserve the seeds with sugar or salt for
their wives to eat and also enjoy pleasure. (While I was a student at
Cracow,[19] from hunger we often ate pottage made from the powder
of Cannabis seed; never, I recall, did it intoxicate or fill me with joy.)
Pollen shaken from flowerets is immediately sifted through a fine linen
cloth. The siftings are at once made into troches. What passes through
with the first and rather gentle motion is called Tsjersi Gobaar. It is
the finest and is reserved for internal use. One pound of it sells for five
ounces of silver. Its excellence is recognized by the sparkling silvery
points that appear in the cracks of the troches. What comes through
on the second shaking of the sieve is called Tsjersi Kamer and is used
externally to enhance the powers of tobacco. A small amount, the size
of a pea, is placed on burning tobacco and inhaled. What comes
through on the last and vigorous shaking is cheaper and green with a
heavier odor and serves the same use. The pollen is made into troches,
or lozenges, with human spittle. The heat of burning coal causes it to
soften, in which state it is more easily formed into a lump. If it smokes
quickly but for just a short time, good quality is indicated; if it smokes
slowly but for a long time, inferior quality is indicated. The leaves are
used for infusion with cold water. This drink suffuses a person with
joys not short of heavy intoxication.

I will describe the manner of preparation by four dervishes (religious mendicants) as I witnessed it at a hostel near the Indian border.
They carried partially crushed leaves and thin vessels necessary for the
preparation in their thigh coverings. They took about two handfuls

of these leaves, which they stirred in water. They pressed the leaves gently and then threw out the water, which had been tainted with the virulent crudity. Fresh water was added, stirred several times with the powder and again thrown out. After washing and pressing the powder twice, they subjected it to a thick wooden pestle and ground it very hard in an uncoated clay vessel until it was reduced to a thick pap. On this they slowly poured fresh water, turned it constantly, and passed it through a linen cloth. Except for waste matter, the entire substance came through, and the purified liquid itself turned green. The person who prepared the liquid measured out a cup containing one-half pound for each of his friends to empty at their pleasure. All the while, he kept turning the vessel so that the powder might not settle to the bottom. When they were filled with the liquid, as one would be with wine, and made joyous, they returned to their travels. Some people mix the powder with syrup to make cakes or lumps to be eaten for the same purpose. In Persia and India men given to intoxicating drugs are commonly called Bengi from the leaves of the Cannabis as a more powerful species of intoxicants. Pliny reports on magic theombrotion, which kings have customarily eaten and drunk.[20] Alessandro Alessandri (*Dies Geniales* 3. 9) related that the abstemious Scythians formerly intoxicated themselves with the smoke of an herb and that the Thracians intoxicated themselves with the fumes of a seed cast into a fire.[21] Theombrotion, the herb, and the seed are no doubt Tsjers, Baeng, and Sjadoneh.

III
Indian Intoxicants

In farther Asia, called India, and stretching as far as China, various drugs are employed to exhilarate, intoxicate, or brutalize the spirits. Among the foreign ones, aside from Tobacco, the most common and the one occupying the mouths of all people is a certain type of apophlegmatic called by the Indians Pinang and Betel after the species employed.[22] It is composed of leaf, fruit, and calx. The leaf, named after the whole plant, is called betel by the Malabars and Siri by the Malays. It has an aromatic savor from mint, the odor of malabathrum or the leaf of the wild canella, but the shape of the leaf of the pepper and cubeb, the stem of which plants has the nature and appearance of a climber. Except for fruit, which it rarely bears, it resembles the long pepper, has an extremely aromatic savor and is sought after, when it can be had, for this composition. The fruit is a nut, or rather a kernel, called Areek and Pinang by the Indians and Fufel and Faufel by the Persians and Arabs. It is similar in size and variegated coloring to the nutmeg. It has a lignose substance capable of disturbing the brain and is very greatly astringent. It grows on a certain palm whose description is found in botanical works. The calx is not processed from mineral stone but from the burned shells of ostrea and conch that the ocean deposits upon the shore.

From these three elements a small mouthful, which physicians call a masticatory, is made in the following manner. Approximately one-third of an areca nut, cut into thin slices or bits, is placed in a (previously deveined) betel leaf rolled into the form of a troche. Then the inner side of the leaf is smeared with a tiny bit of calx weighing about one-half a scruple.[23] This done, the troche is closed to prevent the contents from escaping, and a lump is created large enough to fill approximately the mouth. As this is rolled and worn down with the teeth, it elicits abundant saliva, which it dyes blood red. Meanwhile, it lends strength to the gums, a purple color to the lips, and a most agreeable breath to the mouth, and what is of prime importance, it gently soothes and exhilarates the spirits by intoxicating the brain. All the Indians, expecially the dark ones in the tropics, and not excluding foreigners residing there, are strangely accustomed and addicted to its use. Some people do not leave home unless properly furnished with troches in their loincloth or a pocket of their garment; they stop chewing only to eat or sleep. Hence, girls raised in softness and idleness spend the whole day, the plebeian girls the entire morning, making betels in order to have them in readiness both for themselves to enjoy slowly and to offer guests and acquaintances according to the custom of the nation. Persons of rank more than other people wantonly idle away time on mastication (pardon the use of the word). As a result their mouths constantly flow with a redness—a disgusting sight! At the same time, it is difficult to understand their speech when the tongue is occupied with another movement.

At the court of Malabar I saw an individual who was standing close to His Highness repeatedly offer fresh betels, not with his fingers but with a thin fork plated with silver and adorned with an array of troches in a long row. This manner of display was designed to enhance the splendor of the court. It is the custom at the court of Siam[24] that members of the king's council (they are called Opra) be presented by His Majesty with containers for betels artistically wrought from gold, silver, or a less expensive metal according to the dignity of the individual. They would not presume to appear at the royal palace or a gathering of the illustrious without this distinguishing mark of nobility. The case is adorned with a small box filled with calx prepared to the consistency of an ointment, a spatula to manage the calx, forceps of a unique design with which the rather hard areca nut can be cut without difficulty, and small compartments for leaves, nuts, and the rolls or little troches made from these. This region secures the leaves from elsewhere, and while the leaves are termed flaccid, they nevertheless have undiminished power and are suitable for extended preservation. For want of these leaves, some people substitute mint leaves, which as nearly as possible furnish their savor. Of these three elements: the leaf contributes grace of savor and odor to the compound; the fruit disturbs and intoxicates the brain with delight; the added calx draws out strength and tempers vehemence; the sharpness of the calx itself, however, is blunted by the astringent power of the nut, for otherwise it would wretchedly gnaw at the tongue and palate. The reduction of the quantity of nut added cannot prevent the tongue of the constant chewer from eventually growing callous, the sense of taste from being destroyed, and the very teeth, which it otherwise usually strengthens, from eroding. There exist many varieties of areca, and among those of lesser potency, the more tender and fresher ones are said to disturb the brain less. For my part I was never able, whatever species I received, to chew a betel without experiencing constant anxiety, cold sweat, and vertigo, the same as produced by tobacco smoke.

I have spoken about opium and its common use by the Persians and Indians. I add an abominable abuse that thrives among the dark Indians: brutalizing the spirits in order to acquire the audacity to perpetrate murder. When tired of their own life or injuries, they dedicate themselves to seeking death through the punishment and death of others. For this purpose they swallow a lump of opium from which the intention conceived is stimulated, reason is disturbed, and the spirit is freed from restraint. The result is such that with dagger in hand they rush like raging tigers into the streets and are bent on slaughtering whomever they meet, whether friend or foe, until they themselves are struck and laid low by someone else. This act, a frequent sight among the inhabitants of Java and the farther Orient, is called Hamuk.[25] The very sound of the word terrifies anyone there who hears it. For if a person sees a murderer, he cries out at the top of his voice the word Hamuk in order to warn those unarmed to flee and take care for their lives, while whoever is armed and courageous is expected to run and slay the beast. There is also an external use of opium among the dark Indians: they steep tobacco with opium diluted in water so that when smoked it more violently disturbs the head. In Java I saw flimsy shops made of reed in which this type of smoking tobacco was displayed for passers-by. No product sold by the Dutch throughout India offers a greater profit than Afiuun, for addicts can neither do without it nor acquire it except when it is transported in Dutch ships from Bengal and Coromandel.[26]

Among the intoxicants employed by the wealthier Indians certain electuaries that work utter magic would each fill a page. Their descriptions include Dutr or *Datura* seed,[27] poppy juice (or Afiuun), Tsjers Cannabis powder, and other elements selected from among poisonous plants according to the particulur effect the medicine man wishes to produce. Aromatics are added to correct the wildness and strengthen the chewer's vital spirits. There are many reports of the drug's astonishing ability to produce diverse fantasies at will. But these are in part incredible and unreal. Here is such a one: that by administering a drug an adulterer can so blind a husband that he may with impunity climb into the man's bed while he is present. It is reported by the most respected Mandelslo[28] and Linschoten.[29] During my stay in various realms of India, I did not chance to test the truth of this effect. This, however, I did learn: that within the chewer's brain marvelous phenomena and phantasms can be created that ignorant men easily believed to be magic worked with the aid of a spirit; and such they do indeed believe, for they have no knowledge of the drug administered and its effect. Here I wish to add examples of this, which I learned from personal experience.

There is a festival celebrated in inland Malabar, during which vestal virgins (sacred to the god Vishnu) are presented for possession by the bestower of the gifts of seasons and produce—their idol—and his spirits, so that conciliated by this pleasure he may spare their crops and cattle.[30] The possession is presented for viewing in a public field with this annual ritual. The virgins are led from the temple by an escort of Brahmins. First, they stand still for inspection by the public. They are beautiful and well groomed; their faces suggest modesty and give no indication of insanity. Without delay, as the priest reads aloud the formula of words from the Vedas[31] (pagan sacred books), the girls move lightly, soon dance, and at length tire their bodies with frequent leaps and irregular movements; they twist their limbs and eyes, foam, and perform horrible actions. The public believes that the girls are so impelled by demons within. While this performance is going on, cymbals and tympanies resound with joy as the people show their approval with words and sighs. When the girls have become tired, the Brahmins lead them back into the temple and instruct them to lie down to rest. After scarcely an hour, the girls are restored to their senses and again led before the public to let the pagan crowd see them freed from spirits and believe that the idol has been conciliated. O cheap fraud! Before the first procession, the priests had administered a drink of some maddening opiate through whose power the girls were carried to these preconceived and intended actions, all the more quickly because the rather strenuous movements accelerated the operation and effectiveness of the drug. Upon being led back, the girls were given a drink of a second drug, an antidote, to check the operation of the first. The crowd suspected nothing, because these secret rites are conducted in the shadowy sanctuary illuminated by a single lamp. The sanctuary constitutes the third and last section of a temple (the structure of their temples is tripartite). The information that the priests administered drugs was reported to me by an Indo-Dutchman, a man of proven integrity and well acquainted with the language. He had stuck his head in through the small door of the sanctuary and was most attentive witness to the fraud perpetrated within.

I witnessed another case in the same region. On a very tall, movable machine seven noble athletes were suspended by two iron hooks driven into the muscles of the back. They held in one hand a sword and in the other a huge shield so that they could sport with arms in honor of the goddess Rhauni, for this conquering heroine is believed to find the greatest delight in blood and wrestling contests.[32] When the athletes were driven around the temple of this goddess three times without interruption by workmen who were spectators, they brandished their arms like fencing masters. Then they also moved their heads, arms, and legs with a theatrical grace as if they were floating through the air. They showed no indication of pain from their movements or bloody suspension. Constantly forcing my way into the circle that pressed around the suspended athletes, I saw a Brahmin administering a certain drink made with the shell of the coconut, without doubt a narcotic to kill the pain that the people believed was removed by the god presiding over the temple. Upon being questioned by me, the priest replied: the juice is Lansjes administered against thirst and exhaustion. Lansjes is the unripe coconut containing a sweet liquid that in time coagulates into a kernel.

Another instance comes from the experience of myself and my comrades. We six Europeans along with the Director, Lord Ly-kochthon,[33] were received by Bengals at a garden party one mile outside of Gamron. During dinner as we drank to the health of each person, the abstemious Bengals by way of compensation took lumps of an intoxicating electuary recently sent them from their native land. Anxious to taste it, I took a mouthful of no mean size. When I praised its ambrosial quality, Lord Lykochthon himself took some, and the rest followed the example of our Director. Only Lord Verdunck, who had previously experienced its powers, refused.[34] What happened? As many of us as were filled with the milk of *nepenthês*, were suffused and replete with a certain inexplicable joy the like of which I have never experienced in my life. Under the power of the drug we could do nothing but exchange embraces, speak very sparingly, smile at those speaking, and move one another to laughter. When the party ended shortly before night, we mounted our horses. With the change of scene the power of the drug created other fantasies in our brains. It seemed as if we were flying with Pegasus through clouds and rainbows with a brilliant display of the most elegant colors meeting our eyes in every direction. When we arrived home and were requested to sit down to the dinner that awaited us, we each of us took with the appetite of a wolf whatever he found before him. We were so delighted with the taste of every sort of dish that we seemed to be dining with the gods themselves.

On the following day no hangover (the Russians call it Pochmeli), as is usual after intoxication, troubled our heads. When sleep dispelled the vapors, reason was restored to its proper state, and we felt excellent. Although the memory of yesterday's actions was effaced (Verdunck refreshed it), we were not the least forgetful of the joys and incomparable pleasure that enfolded us. We recollected also (which I note as extraordinary!) the fear of rushing toward the right side, which constantly troubled us whether on horse or on foot. The idea so strongly impressed itself that we could not get it out of our minds. Reader, there you have the magic intoxication induced by a drug. If about evening you administer such a drug to a melancholiac, upon awakening he will believe and will assert under torture that he has flown through the air around the Brocken specter,[35] even though he must face a death sentence from the judges.

Notes

1. Coco flower is the flower of the coconut palm (*Cocos nucifera*) whose sweet juice ferments readily and is used to make the strong alcoholic drink *arrack*.

2. Nicotine is a poisonous, colorless, oily alkaloid with a pungent odor and acrid taste. It is found in the leaves of the tobacco plant in combination with citric and malic acids.

3. White poppy (*Papaver somniferum*) is the source of opium.

4. *Cannabis sativa* furnishes hashish and marijuana.

5. Satyrion is any of the various types of *Orchis* (orchids) and is reputed to be an aphrodisiac.

6. Tobacco: from the tube or pipe in which the Indians smoked the plant. The Spaniards transferred the name to the plant itself.

7. Kaempfer is referring to the island of Tobago, north of Trinidad, visited by Columbus on his third voyage to the Americas in 1498. There is no indication, however, that tobacco was grown on Tobago.

8. *Hyoscyamus*: a genus of poisonous herbs yielding a poisonous crystalline alkaloid, hyoscyamine, which is sedative and antispasmodic.

9. Francesco Redi (1626–97) was the author of *Osservazioni intorno alle vipere* (Florence, 1664), the first comprehensive description of snake poison. Redi's experiments with snake venom also made use of tobacco oil to show that venom must be injected to be effective.

10. Located close to the Volga, Kazan became the capital of an independent Tartar khanate founded in 1438 but was captured by Ivan the Terrible in 1552.

11. Khaliaan or Khaliuun: Persian term for hookah.

12. The antidote is named for Andromache, who survived Hector, because her many years of suffering and enslavement could only be tolerated by a person whose serenity was comparable to the effects of theriac.

13. Trunken Krämere: drunken shopkeepers.

14. *mêkonion*: the juice of the poppy.

15. *nepenthês*: the Greek word means "banishing pain and sorrow." A *pharmakon nepenthês*, a drug, perhaps opium, is given to Helen of Troy by an Egyptian:

Then Helen, daughter of Zeus, took another counsel. Immediately she cast into the wine, which they were drinking, a drug to quiet all pain and strife, and to bring forgetfulness of every ill. Whoever might drink this down, when it is mingled in the bowl, would not in the course of that day let a tear fall down over his cheeks, no, not though his mother and father should lie there dead, or though before his face men should slay with the sword his brother or dear son, and his own eyes beheld the sight. (*Odyssey* 4. 219-26)

16. Bangue (*bhang*) is the East Indian name for the hemp plant, but the term is now used for a narcotic prepared from the plant.

17. Hendrik Adrian van Reede tot Draakestein, *Hortus Indicus Malabaricus*, 12 vols. (Amsterdam, 1678–1703).

18. Luristan, "Land of the Lurs," a largely mountainous region of western Persia. Located just to the west of Shiraz, Kazerun lies in the province of Fars, which borders on the Persian Gulf.

19. Kaempfer studied languages, medicine, and philosophy at Cracow during the years 1676–80.

20. Pliny observes, "the theombrotion ["food for the gods"] grows . . . being like a peacock in its colorings and of a very fine scent . . . the kings of Persia take it in drink for all bodily disorders and to keep unshaken their intellectual powers and sense of justice; it is also called semnion (the august plant) from the majesty of its powers" (*Historia Naturalis* 24.102.162).

21. Alessandro Alessandri (Alexander ab Alexandro, 1461–1523) abandoned the practice of law to devote himself to literary studies at Rome. Modeled on the *Attic Nights* of Aulus Gellius, the *Saturnalia* of Macrobius, and the *Policraticus* of John of Salisbury, his *Alexandri de Alexandro Dies Geniales* was first published at Rome in 1522. This work has been compared with a large department store offering a wide variety of merchandise, some useless, some doubtful, but some precious—selection must be made by an experienced shopper.

22. The species is the betel palm (*Areca catechu*).

23. scruple: an apothecary weight of twenty grains, the third part of a dram, or 1/24 ounce.

24. On the voyage from Batavia to Nagasaki in 1690, Kaempfer visited Ayuthia, the capital of Siam.

25. Hamuk (Malay, *amok*): to go crazy, to run amok. Kaempfer reached Java in September 1689, where he spent eight months before sailing to Nagasaki. "At the cry of 'Amok', anyone could kill the native," reported C. P. Thunberg

(*Travels in Europe, Africa, and Asia, Made Between the Years 1770 and 1779*, 3rd ed. [London, 1795] 2:286).

26. Coromandel: the coast of southeast India stretching north from Cape Comorin at the tip of India to the Kistna River.

27. *Datura*: a genus of poisonous plants of several species which belong to the potato family.

28. Johann Albrecht von Mandelslo. *The Voyages and Travels of J. A. de Mandelslo into the East Indies, Begun in the year 1638 and finish'd in 1640*, trans. J. Davies (London, 1662).

29. Jan Huyghen van Linschoten (c. 1563–1611) traveled to Goa and participated in the Dutch arctic voyages. His influential *Navigatio ac itinerarium* (1595–96), which includes his own experiences as well as the reports of other travelers, was translated into English, German, and French. Kaempfer's reference can be found on pp. 209–10 of the English (1598) translation (reprinted in 1885 by the Hakluyt Society).

30. In Hindu theology the trinity is composed of Vishnu, the Preserver, along with Brahma and Siva.

31. Vedas: the oldest sacred literature of Hinduism.

32. Rhauni is the wife of Siva, usually named Kali, and is associated with death and destruction.

33. Vice-Admiral Wibrand Lykochthon, commander of the Dutch East India squadron in which Kaempfer sailed from Gamron (now Bandar Abbas) in Persia to Java.

34. Verdunck was an official of the Dutch East India Company.

35. Brocken is the highest peak in the Harz Mountains. The specter is a greatly magnified human shadow cast upon a cloudbank when the sun is low. It follows every motion of the observer in the form of a huge but misty image of himself.

Observation XVI
Magic Spells of the Makassars,
or Inhabitants of Celebes

I
Superstitious and Inane Spells of Various Nations

I offer a second observation concerning the native magic of the Indies.[1] While it may be less than suitable for chaste ears, it is nevertheless not unworthy of contemplation. My subject is the binding spells of the Makassars, which are lauded by the people of the Indies. With these spells they enervate a person's virility or revive it from the state of enervation. Vergil's eighth eclogue, which illustrates the art of enervation, makes it clear that this magic is not of recent origin: "Weave three colors with three knots, Amaryllis, / Weave, Amaryllis, but say, I weave the chains of Love."[2] The wise men of all ancient peoples profess their belief in its efficacy. In noteworthy examples they show no hesitation to believe natural functions are bolstered through the aid of the devil. Moreover, a later age, enlightened by a purer religion, has not so renounced slavish credulity and baseless superstition as to refrain from augmenting every faculty of man. In other areas men may furnish the world with documents of unsullied judgment and superb erudition, yet when engaged in this genre of argument, they do not blush to cull the remains of credulous antiquity and to cite numerous cases from them. I take the liberty to point out as outstanding examples Del Rio,[3] Arnisaeus,[4] Hieronymus Jordan,[5] and Rodrigues De Castro.[6]

But let us return to the Makassars. This nation possesses a spirit unbroken in war, gravity in conversation, and a noble nature. Such endowments are complemented by a dignified complexion and beauty that earn this nation the reputation as the most noble of the dark skinned. The Makassars praise nothing more highly than reciprocal and constant trust toward the trustworthy. They also condemn nothing more severely than the secret and persistent lust for vengeance against the untrustworthy and treacherous, which is concealed until gratified. Consequently, the weaker sex never dismisses husbands or lovers caught betraying their love without inducing a spell as punishment. Quite nobly, the ladies cause no harm to life or vital organs, but impair solely the instrument of treachery in such a way the gentleman is not capable of intercourse with anyone other than his wife or the one-and-only to whom he has pledged his faith. The gentlemen in turn and with like retribution so restrain adulterers who connive at intercourse with their women that they grow ingloriously limp on every attempt. This practice of casting sexual spells appears to have spread from Makassar to the surrounding nations, especially (so far as I am aware) to Java, Malaya, and Siam.[7] These nations share the common tradition (from petulance in the case of suitors and concubines or as a punishment in the case of unfaithful husbands and wives) of casting a spell and rendering the sex act impossible with others.

The art of casting spells flourishes, perhaps in an even more superstitious form, among the Christians of the East. These Christians are firmly persuaded that during the church ceremonies and especially at the very moment of marriage, a debility can be inflicted upon the bridegroom and that its inducement can be averted through magic. Their fear reaches such proportions that many of the nobles and the wealthy are careful to have a witch present at the marriage service. The witch is supposed to know how to frustrate a binding spell and consequently how to avert from the newlyweds the marital disagreements that may arise from this source. I can cite many examples that I heard from the principals themselves as well as those I learned of from being present at weddings. During my stay at the Court of Russia, a certain individual, a friend of mine (a Dumdiake)[8] closely connected with His Royal Highness, was to celebrate his marriage to an illustrious maiden. As a matter of prime importance and upon the urging of both families, he saw to it that there was present an old woman whose magic powers were renowned among the residents of Sloboda (a quarter inhabited by Germans).[9] Her function was to thwart by her own art any attempt to cast a spell that might be made by someone present for the religious ceremonies. My friend reported that this practice was both usual and licit for weddings of the nobility.

I was not in the least shocked, since the czar and the patriarch, who ought to prohibit superstitions, were themselves wrapped in the darkness of superstition and firmly spellbound.

We were invited, together with the ambassador, to attend a solemn wedding ceremony held in the evening at the principal church of the Armenian suburb of Isfahan.[10] Quite by chance, we saw a man casting a spell as he hid behind the door of the church. While muttering, he cut into small pieces a long rod that he was holding under his armpit. Leuenanker, the distinguished and noble Swede,[11] was the first to notice the actions of the man. He asked the man what his purpose was and admonished him about the need for reverence in church. The man gave no answer and continued his work. At length a relative approached us after he had humbly prayed to avert impediments to marriage. He explained that the man's actions were undertaken by order of the bride and groom and were designed to render ineffective the designs of any practitioner of magic spells who might be among the guests.

While the public performance of these superstitious rites is licit for the Christians of the East, our own people, the Christians of the West, do in fact secretly engage in them despite legal and religious prohibitions against the poison of superstition. Nor is support from philosophers lacking. They contend that binding spells can be induced by locking a door, tightening a strap, implanting a knife in a wall, or other acts undertaken at the moment of holy marriage. They openly teach that by the principle of countering stratagem with stratagem, binding spells induced for such a purpose can be overcome by urinating through a wedding ring or the handle on the door of a stone sepulcher and by similar absurd phylacteries that do not rely on natural forces.

Such pronouncements are unworthy of philosophers! For this is an infallible law of truth, as Sennert says in his *Hypom. Phys.*:[12]

> Nothing acts beyond the powers of its own species. Similarly, it is wrong to assert that bindings, locking of doors, and other actions or words muttered by witches can induce impotence. Exception may be made in the case of a man of superstitious bent, who through the workings of his own imagination might easily impede or promote the flow of spirits needed to enlarge the penis.

In the case of newlyweds, who are often bound and restrained from their marital function by a sense of shame, experience and familiarity will loosen and restore their sexual powers.

Authors who try to explain effects of this type as generated by the intention and mental power of a spellcaster directed against the person or body of another do not properly distinguish between permanent and temporary conditions. Although Francis Bacon [13] and van Helmont [14] are the chief proponents, the thesis is expounded with absurdities that must be rejected.

II
Binding Spells of the Makassars

Perhaps a somewhat milder judgment should be passed on the types of binding spells originating among the Indians and the Makassars. Not only are these spells performed with harmless and ineffectual actions, but additionally the means and methods employed are natural. By magnetic flux and harmony, a type of mental estrangement is produced in the adulterer; and by an additional slight impulse of the consciousness and imagination, the vital spirits, those highly fluid particles even more inconstant than the human will, are recalled from a lapse. In my opinion, one could even argue that a limp penis can naturally be induced in an instant by some other process quite obscure to me, since a ready supply of examples hardly permits doubt.

I have listened to many humorous stories about spells, but I considered them the remarks of idlers and unworthy of credence. At length by a stroke of good fortune, I met a Dutch military officer, a respectable and athletic young man. He was a fellow countryman of mine participating in the expedition against the rebel leader, whom the Dutch called Captain Jonker, [15] on the island of Java Major. As I enjoyed his trust, he told me the following. He had returned to Holland from Makassar two years ago and thereupon married a Dutch girl, the niece of the Treasurer of the Dutch East India Company at Amsterdam, Bewind hebbers Nichte. When he attempted to engage in intercourse (as he had been accustomed to before his departure), his penis went limp, though he was a stout warrior on this field of battle. He recalled his lover at Makassar with whom he had lived during his tour of duty as if they were husband and wife—such is the custom in that land. Before his departure, he betrayed her by making love to another woman. Now he readily understood that his lover had put a spell on him as a punishment before his departure. Upon the recommendation of his patron and with the permission of the governing board, he returned to the Indies to revisit the place where he had been stripped of his virility. Because his lover had died, he fully regained his virility but only with great difficulty. With his virility restored, he planned to return to his native land.

When I related this brief tale, which seemed to be both fascinating and novel, at a gathering of illustrious gentlemen in Batavia, one of them, G. D., commented that he was familiar with such cases. He assured me that I should entertain no doubts about the truth of this story, since upon recommendation many men must return to their former residences in order to restore potency to their wasted members. Subsequently, I was more alert to such reports, and everywhere I discovered an abundance of examples to support this position. I have no desire here to present myself as a collector and distributor of such stories.

In conclusion, I wish to add as a final flourish the occult technique itself of sexual binding and loosening. By money and entreaties, I extracted it from an old witch who was a practitioner of this medicine. She said that there are many methods of sexual binding but that they scarcely compare with her own secret method that had been tried and tested by long experience. These are the old woman's words:

The girl who wishes to bind her lover, or the wife who wishes to bind her husband, must with an undergarment or any cloth wipe clean his penis after the sex act, attempting to gather up as much semen as possible. She is to fold the cloth properly and bury it in the earth beneath the threshold of her home. For as long as the cloth remains buried there, so long will his penis obey the will of no one but the woman who casts the spell. Nor will the man be freed from this bond before the cloth itself is freed from confinement beneath the threshold. On the other hand, if a man wishes to bind his bed companion, he must burn an undergarment of hers stained with menstrual matter or any cloth well stained with menstrual flux. He must knead the ashes with his own urine and shape the image of a penis. If the ashes are insufficient for making the small image, he must knead them with some earth on which he has recently urinated. He must carefully mold the image and then let it dry. It must be stored in a dry place and protected from all moisture. For as long as he thus preserves this model, every male shaft directed at his woman will instantly droop. But when the spellcaster himself is ready to enjoy his longed-for pleasure, he must first moisten this magic model. For as long as it is wet, the spell will be suspended and his penis will obey him—but the same is true for all other men whom the woman allows to have sexual intercourse with her.

So much did the old woman say. But what strange magnetic power and relationship lie concealed in this? Let those wiser than I decide! I have merely included this observation for the pleasure that such variety affords the reader.

Notes

1. The first is Observation XV, on Persian and Indian intoxicants.

2. Vergil (70–19 B.C.), *Eclogues* 8. 77–78. References to sexual and other types of magic are common in classical antiquity. Vergil's friend, the poet Horace (65-8 B.C.), has left us a number of examples; see especially *Epodes* 5 and 7. For a brief account of the magic of Greece and Rome, with bibliography, see *The Oxford Classical Dictionary*, 2nd ed. (Oxford, 1970), 637–38.

3. Martinus Antonius Del Rio (1551–1608). Celebrated for his sound scholarship, Del Rio, at the age of 19, wrote a commentary on the tragedies of Seneca in which about 1,100 authors are cited. He is said to have read ten languages and every book worth reading. In 1580 he joined the Society of Jesus. His *Disquisitionum Magicarum* (Louvain, 1599–1601), a credulous analysis of and implacable attack on witchcraft, was recognized as the standard treatise on the subject.

4. Henningus Arnisaeus, a German physician who wrote on medicine, politics, and jurisprudence. Among his many works is *Observationes Anatomicae* (Frankfurt, 1610). Arnisaeus (born c. 1580) died in 1636 while serving as first physician to the king of Denmark, Christian IV.

5. Hieronymus Jordan, a physician who wrote on nutrition, paralysis, and salt. See *De eo, quod Divinum, aut supernaturale est in morbis humani corporis, ejusque curatione liber* (Frankfurt, 1651).

6. Estevam Rodrigues De Castro (1559–1638), Portuguese poet and physician who contributed to gynecology. Among his numerous works is *Opusculum de mutatione aliorum morborum in aliis, in quatuor libros divisum, medicinae studiosis valde utile et recondita doctrina refertum* (Florence, 1627).

7. See Walter William Skeat, *Malay Magic, An Introduction to the Folklore and Popular Religion of the Malay Peninsular* (London, 1900).

8. For this period of Kaempfer's travels (7 July to 5 September 1683), see Karl Meier-Lemgo, *Die Reisetagebücher Engelbert Kaempfers* (Weisbaden, 1968), 21-28. *Dumnyj d'jak* is the title of a person who served as head of various government departments of Russia.

9. The name of the quarter transliterates from the Russian as *Sloboda* and is pronounced *slabada*.

10. The suburb is called Sulpha, or Sjulfa, and is located to the south of the city of Isfahan across the river. Kaempfer reports that Sulpha was inhabited by

30,000 schismatic Armenian Christians and had 13 churches (*Amoenitates Exoticae*, 164). After voyaging from Russia over the Caspian Sea, Kaempfer remained in Persia from 22 November 1683 to 30 June 1688, when he sailed for India.

11. Leuenanker is identified by Kaempfer as one of the members of the Swedish embassy to the court at Isfahan, and as the son of the head director of Swedish trade (*Amoenitates Exoticae*, 221).

12. Daniel Sennert (1572–1637), *Physica hypomnemata. I. De rerum naturalium principiis. II. De ocultis qualitatibus. III. De atomis & mistione. IV. De generatione viuentium. V. De spontaneo viuentium ortu* (Lyons, 1637), also published as *Hypomnemata physica* (Frankfurt, 1636). In Observation IV, Kaempfer calls Sennert the "German Galen."

13. Sir Francis Bacon (1561–1626), English philosopher, essayist, and lord chancellor. Bacon says, "If a witch by imagination should hurt any afar off, it cannot be naturally; but by working upon the spirit of some that cometh to the witch; and from that party upon the imagination of another; and so upon another; till it come to one that hath resort to the party intended; and so by him to the party intended himself" (*Sylva Sylvarum* cent. x, n. 950).

14. Jean-Baptiste van Helmont (1577–1644), a Belgian chemist and iatrochemical physician, Helmont both contributed to scientific progress and fostered murky superstition. While credited with being the first to distinguish gases (the word *gas* is his creation) from atmospheric air, he believed that certain occurrences in the body were brought about by the devil through witches as intermediaries. Helmont's works, *Ortus medicinae, vel opera et opuscula omnia* were published by his son, Franz Mercurius (Amsterdam, 1688).

15. Captain Jonker (1630–1689), a full-blood Amboinese, commanded auxiliaries in support of the Dutch. Jonker, however, shared in a conspiracy against the Dutch. When his arrest was ordered, Jonker took up arms against the government and met his death in battle.

Select Bibliography

Index

Select Bibliography

Aitchison, James Edward Tierney. "Notes to Assist a Further Knowledge of the Products of Western Afghanistan and of North-Eastern Persia." *Transactions of the Botanical Society, Edinburgh* 18 (1891): 1–288.

Banks, Sir Joseph. "An Account of Ambergris, by Dr. Schwediawer; presented by Sir Joseph Banks, P.R.S." *Philosophical Transactions of the Royal Society of London* 73 (1783): 226–41.

Bauhin, Jean. *Historia plantarum universalis nova et absolutissima.* Yverdon, 1650–51.

Bowers, John Z., and Robert W. Carrubba. "The Doctoral Thesis of Engelbert Kaempfer on Tropical Diseases, Oriental Medicine, and Exotic Natural Phenomena." *Journal of the History of Medicine and Allied Sciences* 25 (1970): 270–310.

———. "Drug Abuse and Sexual Binding Spells in Seventeenth Century Asia: Essays from the Amoenitatum Exoticarum of Engelbert Kaempfer." *Journal of the History of Medicine and Allied Sciences* 33 (1978): 318–43.

———. "Exotic Antidotes in Seventeenth-Century Asia." *Clio Medica: Acta Academiae Internationalis Historiae Medicinae* 14 (1980): 129–40.

Carrubba, Robert W. "Seventeenth-Century Latin Accounts of Acupuncture and Moxibustion." *Acta Orientalia* 40 (1979): 205–27.

———. "A Seventeenth–Century Report on the Snake Charmers of India." *Acta: Fourth International Congress of Neo-Latin Studies* (1985): 18–26.

———. "Kaempfer's Latin Account of the Magic of the Makassars." *Acta Orientalia* 49 (1988): 37–47.

———. "Engelbert Kaempfer and the Myth of the Whirlpools of the Caspian Sea." *Proceedings of the Seventh International Congress of Neo–Latin Studies* (1991): 261–71.

———. "Engelbert Kaempfer and the Myth of the Scythian Lamb." *Classical World* 87 (1993): 41–47.

Carrubba, Robert W., and John Z. Bowers. "The Western World's First Detailed Treatise on Acupuncture: Willem Ten Rhijne's *De Acupunctura.*" *Journal of the History of Medicine and Allied Sciences* 29 (1974): 371–98.

———. "The First Report of the Electric Torpedo Fish of the Persian Gulf." *Journal of the History of Biology* 15 (1982): 263–74.

Charas, Moses. *New Experiments on Vipers.* London, 1670.

da Orta, Garcia. *Coloquios dos simples, e drogas he cousas medicinais da India.* Goa, 1563.

Dawson, Warren R. "Mummy as a Drug." *Proceedings of the Royal Society of Medicine* 21 (1927): 34–39.

de Bondt, Jakob. *De medicina Indorum lib. IV.* Leiden, 1642.

de l'Escluse, Charles. *Exoticorum libri decem.* Antwerp, 1605.

Del Rio, Martinus Antonius. *Disquisitionum Magicarum.* Louvain, 1599–1601.

Grundfest, Harry. "Electric Fishes." *Scientific American* 203 (1960): 115–24, 220.

Hüls, Hans, and Hans Hoppe. *Engelbert Kaempfer zum 330 Geburtstag.* Lemgo, 1982.

Idyll, C. P. "Ambergris—Neptune's Treasure." *Annual Report of the Board of Regents of the Smithsonian Instituion 1959.* Washington, 1960.

Kaempfer, Engelbert. *Exercitatio Politica de Majestatis Divisione.* Danzig, 1673.

———. *Disputatio medica inauguralis exhibens decadem observationum exoticarum.* Leyden, 1694.

———. *Amoenitatum exoticarum politico-physico medicarum fasciculi V, quibus continentur variae relationes, observationes & descriptiones rerum Persicarum & Ulterioris Asiae.* Lemgo, 1712.

———. *The History of Japan.* Translated by J. G. Scheuchzer. 2 vols. London, 1727.

Kircher, Athanasius. *De arte magnetica.* 3rd ed. Rome, 1654.

Lee, Henry. *The Vegetable Lamb of Tartary: A Curious Fable of the Cotton Plant.* London, 1887.

Ludolf, Hiob. *Historia Aethiopica.* Frankfurt, 1681.

Mattioli, Pietro Andrea. *Commentarii in libros sex Pedacii Dioscoridis Anazarbei de medica materia, Opera quae extant omnia.* Frankfurt, 1598.

Meier-Lemgo, Karl. *Die Reisetagebücher Engelbert Kaempfers.* Weisbaden, 1968.

Mercuriale, Jerome. *De Morbis cutaneis et omnibus corporis humani excretionibus.* Venice, 1572.

Paré, Ambroise. *Discours de la Mumie et de la Licôrne.* Paris, 1582.

Ray, John. *Historia Plantarum.* 3 vols. London, 1686–1704.

———. *Synopsis Methodica Animalium Quadrupedum et Serpentini Generis.* London, 1693.

Redi, Francesco. *Osservazioni intorno alle vipere.* Florence, 1664.

————. *Opuscoli varj di F. Redi.* Florence. 1684–91.

Rumpf, Georg Eberhard. *Herbarium Amboinense.* 6 vols. Amsterdam, 1741–50.

Scaliger, Julius Caesar. *Exotericarum exercitationum liber XV.* Frankfurt, 1557.

Sennert, Daniel. *Physica hypomnemata. I. De rerum naturalium principiis. II. De ocultis qualitatibus. III. De atomis & mistione. IV. De generatione viuentium. V. De spontaneo viuentium ortu.* Lyons, 1637.

Severino, Marcus Aurelius. *De abscessuum recondita natura libri VIII.* Naples, 1632.

Sollmann, Torald. "A Sketch of the Medical History of Dragon's Blood." *Journal of The American Pharmaceutical Association* 9 (1920): 141–44.

Stearn, William T. "Botanical Exploration to the Time of Linnaeus." *Proceedings of the Linnean Society of London* 169 (1958): 173–96.

ten Rhyne, Wilhem. *Dissertatio de arthritide: mantissa schematica: de acupunctura: et orationes tres I. de chymiae et botaniae antiquitate et dignitate. II. de physionomia. III. de monstris.* London, 1683.

Valentini, Michael Bernhard. *Museum museorum.* Frankfurt, 1704.

van Reede tot Draakestein, Hendrik Adrian. *Hortus Indicus Malabaricus.* 12 vols. Amsterdam, 1678–1703.

Velschius, Georgius Hieronymus. *Exercitatio de vena medinensi, ad mentem Ebnsinae, sive de dracunculis veterum.* Augsburg, 1674.

von Herberstein, Sigismund. *Rerum muscoviticarum commentarii.* Vienna, 1549.

von Mandelslo, Johann Albrecht. *The Voyages and Travels of J. A. de Mandelslo into the East Indies. Begun in the year 1638 and finish'd in 1640.* London, 1662.

Worm, Ole. *Musaeum Wormianum.* Leiden, 1665.

Index

acupuncture, xviii, 108-19; development, 117n.4; Faritatte, 110; Fineri barri, 113; Fuda barri, 113; Gecquan, 114; Kawa kiri, 124; needles, 109-13; as preventative, 124-25; procedure, 113-16; Sjoquan, 114; Tensasj, 110; Tsjuquan, 114; Uuts barri, 113

Aitchison, J. E. T., 63n.24

Alessandri, Alessandro, 189, 197n.21

aloeswood, 128, 138n.22

amber (Succinum), 174-75

ambergris, xviii, 170-79; adulteration, 172-73; amber (Succinum) and, 174-75; *Ambra subalbida,* 173; attributes, 172-74; bees and, 170, 171; buskoppe fish, 174; composition, 176n.1; Kusura no fuu, 173; medicinal uses, 174, 175, 178-79n.23; perfume and, 177n.12; tests for authenticity, 173, 177n.14; theories, 170-71; whales and, 173, 174, 177-78n.15

Amboina, 69, 71n.11

Amoenitates Exoticae (Kaempfer). *See Exotic Pleasures*

Andrum (Andu waja ku). *See* hydrocele

antidotes: for colic, 115; for hydrocele, 75; mercury, 75; moringa, 75, 76n.11; for Pericàl, 79-81; snake stone, 102-5; theriac, 105, 106-7n.13, 185

Apes of Mangatti, 74

Arabia, 120, 121, 184

Arbor Alexipharmaca A. Cost. See Moringa

argil, 156, 168n.15

Aristotle, 12, 14; equivocal generation and, 36, 37, 42n.13

Arnisaeus, Henningus, 199, 205n.4

Artemisia vulgaris (latifolia), 80, 110, 117-18nn. 4, 5, 123; Futz, 123; Jamoggi (mugwort), 123, 137n.12; tea preservation and, 158, 168n.17. *See also* moxibustion

arthritis, 125, 160

asafetida (asa). *See* Disguun asafetida

Ausonius, Decimus Magnus, 164, 169n.29

Avicenna (Ebn Sina), 61n.5; on Disguun asafetida, 47; on *dracunculus,* 30, 31, 32; on mummy, 29n.16

Bacon, Francis, 202, 206n.13

Baglivi, Giorgio, 12

Banda Islands, 173, 178n.16

Bandar Abbas. *See* Gamron

Banks, Joseph, 177n.15

Baran (Barreh). *See* Scythian lamb

Barannetz (Borannetz, Barannek). *See* Scythian lamb

Barclay, 4

Bartholin, Casper, 174, 178n.20

Batavia, xvii, 90, 94n.15, 126

bathing: *dracunculus* and, 39-40; moxibustion and, 135

Bauhin, Jean, 1, 69

Baze, William, 94-95n.17

bdellium, 65, 70n.2

Begi, Hasjem, 187

Belessoon. *See* mummy

Bellew, Henry Walter, 63n.24

benzoin, 172, 176-77n.11

Betel (*Areca catechu*), 189-91, 197n.22; names, 189-90; use, 190

bezoar, 65, 70n.1, 107n.14

bitumens, 27n.1. *See also* ambergris; mummy

Blanckaert, Stephen, 37, 42n.14

bone: callus, 25-26, 29n.17; mummy used for healing, 23-26, 29n.17; Pericàl and, 78-79; porotic, 26, 29n.17

Borometz fruit. *See* Scythian lamb

Borrichius, Olaus, 10, 15n.4

Botanic Garden, The (Darwin), 5n.1

Brahmins (Brahmans), 88, 117n.2, 120, 122;

Robert W. Carrubba is provost and vice-chancellor of the University of Wisconsin, Oshkosh. Born in New York City, he received his undergraduate degree from Fordham College and his doctorate from Princeton University. He was professor of Classics and associate dean of the graduate school at Pennsylvania State University and then founding dean of the graduate school at the University of North Carolina at Charlotte. His past professional positions include those of secretary-treasurer of the American Philological Association and president of the Classical Association of the Atlantic States. He has published extensively on the poetry of Horace and on Kaempfer.

THE LIBRARY OF
RENAISSANCE HUMANISM

General Editor
Philip Rollinson
University of South Carolina

European Editor
James W. Binns
University of York

Also in this series

Principles of Letter Writing: A Bilingual Text of
Justi Lipsi Epistolica Institutio
by Justus Lipsius
Edited and Translated by R. V. Young and M. Thomas Hester

Somnium et Vigilia in Somnium Scipionis
(Commentary on the Dream of Scipio)
by Juan Luis Vives
Edited and Translated by Edward V. George
Attic Press, Greenwood, SC

On Poetry
by Giovanni Antonio Viperano
Translated by Philip Rollinson
Attic Press, Greenwood, SC